MAPPING
ENGLAND

MAPPING ENGLAND

SIMON FOXELL

black dog
publishing
london uk

To Anne,

who, while having spent almost her entire life in England,

still claims she is a stranger, while being as English as the rest of us.

INTRODUCTION
08

MAPPING THE NATION
22

USEFUL AND INFORMATIVE
80

IMAGINING ENGLAND
204

Duxton

Chester 111 Lincoln Skegness
109 110 Stoke 112 113 114
Crewe on Trent Nottingham Boston

Derby Grantham King's
118 119 120 121 122 123 124 Lynn 125
Stafford
Shrewsbury Leicester

gomery Birmingham Peterborough
129 130 131 132 133 134 135 136
Worcester Warwick Northampton Bury

Hereford Banbury Bedford Cambridge
142 143 144 145 146 147 148 149
Gloucester Cheltenham Luton Hertford

nmouth Oxford Aylesbury 160 161 Chelms
155 156 157 158 159 162
Bristol Swindon LONDON Sou
Newbury Reading Chatham
165 166 167 168 169 170 171 172
Bath Guildford T Wells
Wells Winchester
Taunton Salisbury 181 182 183 184
177 178 Southampton Portsmouth Brighton Hastir
Lyme Dorchester 179 180 Eastbourne
Regis Bournemouth Isle of Wight
Weymouth

Scale of Index

50   0   50   100 Mile

INTRODUCTION

MAPPING AN IDENTITY

When the monarchy was overthrown by the Commonwealth under Oliver Cromwell in 1649, and after the King—the former embodiment-in-person of the idea of state—had been executed, a new official image became an urgent necessity. Two images were selected for the Great Seal of 1651; Parliament meeting in full session on one side, and on the obverse, a new potent symbol of the nation—a map. This map depicted both England and Ireland, derived from Christopher Saxton's map of 1579 and later versions of the same. The modern idea of a nation as a bordered territory was superseding the idea of the state as personified by a dynastic ruler empowered by divine right. The map, a relatively new arrival as a popular and widely-understood concept, came into its own as the natural expression of that idea.

Any book called *Mapping England* must answer the question "why not 'Mapping Britain' or even 'Mapping the British Isles'? Britain, Great Britain and Ireland (or the British Isles) are indisputably better defined geographic entities, bounded by a non-contestable and well-defined shoreline. If maps and mapping were only about geography, then it would have been the only possible approach, but—although notionally about geography—cartography embraces many other disciplines and offers far greater potential for both creativity and mischief. Maps (and especially those of contested territories, even ones of so comparatively stable a territory as that of Britain) are battlefields of ideas and ideologies; the locus of political, social and cultural skulduggery; the *mise-en-scene* of propaganda, illusion and sleight of hand and, above all, the playground for notions of national identity and definitions of, otherwise imagined, communities.

Large numbers of maps have been published over the last 450 years depicting both the British Isles and England, often in parallel editions. Each time a map has been prepared, printed or purchased, it has demanded a choice: Britain or England. This is, of course, to grossly simplify the options, partly because Wales is almost always included with England (the polity before the mass production of maps) and Britain is normally the full British Isles, with or without outlying islands. This simplification also flattens out the other possible choices for Scotland, Wales and Ireland, with their own traditions of mapmaking, as well as for sub-regions, counties, cities and estates. It is nonetheless a familiar choice for the English, in many fields, and one they have been able to play fast and loose with over the centuries. Where this choice is normally an unspoken one and has provided the capacity for much creative ambiguity (as well as frequent nationalistic and jingoistic *faux pas*), mapmaking requires greater clarity and definition. The mapmakers of the sixteenth century fully defined a place—England—that had been in development as an idea for centuries. This definition came just in time for the question of British place and nationality to become live again with the institution of the dual monarchy, when James VI of Scotland became James I of England after the death of Queen Elizabeth in 1603. The mapmakers who followed were forced to grapple with the consequences, largely opting to offer both choices, a position that meant they carried on producing maps of England, even as the official nation state became the United Kingdom.

This cartographic undermining of the British project towards a United Kingdom perhaps only contributed to the political, legal and tribal divisions of the Kingdom, but while the very disparate regional elements of England cohered within a single identity (as did the very different highland and lowland areas of Scotland within theirs) the overall map of Britain has remained a jigsaw, with very distinct and separate pieces for each of the national parts. There are relatively few lines on the ground that mark the boundaries between these pieces, they only really exist on maps; but their significance and therefore the significance of the maps that define them should not be underestimated, especially as the nations pull apart again and re-assert their own identities.

The maps in this book are a result of the surprisingly persistent idea, one that lasted through the centuries when the British project was at its strongest and most determined, that a map of England was worth producing, investing in and printing, even though the state of England did not exist in any official form, or at least only did so in contradistinction to the other national elements of the United Kingdom. These maps embody an idea of England at a time when, even if it was not officially discouraged, the alternative idea of Britain—and more specifically a Britain taking its rightful place at the heart of a global empire—was being vigorously promoted as the national ideal.

It is because there are ideas at stake being both consciously and unconsciously expressed; of identity, politics, society and, above all, place; that makes this a book about mapping England. That is, unless mapmakers and atlas publishers were simply taking advantage of the fact that England and Wales fit relatively neatly onto a sheet of paper or a page in an upright book, without the greater reduction of scale and detail involved in showing the whole of the British Isles.

THE PRINT REVOLUTION

The Protestant Reformation and—for the purposes of this book—specifically the English reformation under Henry VIII, with its redefinition of an England separate from the mass of Catholic Europe (and Scotland), its own Church and independent sovereignty, coincided with (and was largely made a reality by) the arrival of mechanical printing and the freedom of thought and expression that it precipitated. In the first 40 years following the production of the Gutenberg Bible, approximately 20 million printed volumes were produced across Europe. During the sixteenth century, production swelled to ten times that number and books became available to anyone who was able to read them.[1] In parallel to the outpouring of text-based volumes, the new printing industry also produced the first widely available maps, both in single sheets for wall-

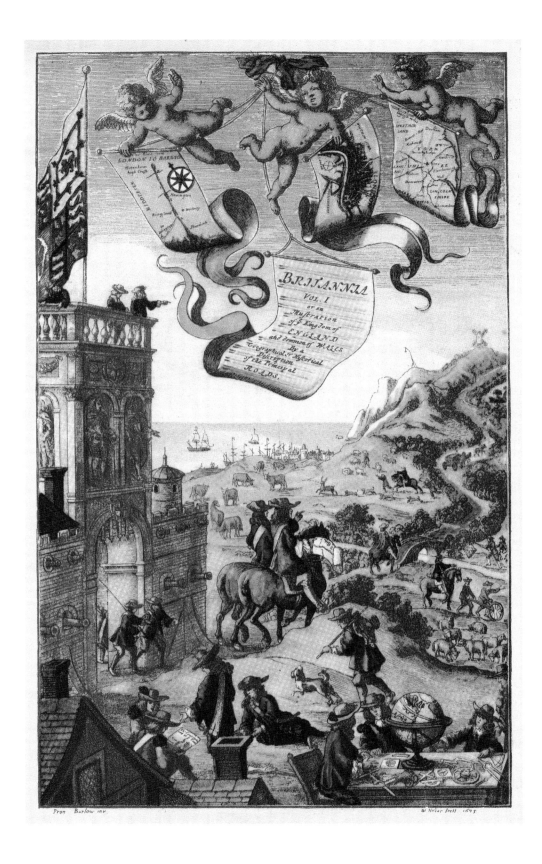

Frontispiece, Britannia, 1675, John Ogilby,
Although Ogilby calls his magnificent book of roads
Britannia, he is careful to subtitle it: *Volume the First: or,*
an Illustration of the Kingdom of England and Dominion
of Wales: By a Geographical and Historical Description
of the Principal Roads thereof. When a further edition
came out in the same year, it became known as the
Itinerarium Angliae.

This frontispiece shows the surveying equipment
and methods Ogilby used: a 'way-wiser', the
handwheel (to measure distance) and a compass to
take triangulation bearings. For centuries, these were
the essential tools for the surveyor and with them
in hand he set out to transform the information
available to all future ws across England.

The Great Seal, 1651

In the wake of the English Civil War and the declaration by the Rump parliament on 19 May 1649 that the country should "henceforth be Governed as a Commonwealth and Free State" new visual expressions of this new state became urgently necessary. This was of particular importance on the new version of the Great Seal of England, released in 1651—the third year of the republic. One side

depicts the House of Commons with the Speaker in the Chair, expressing its new sovereignty and therefore the end of the monarchy, the other side shows a map of England, Wales and Ireland, recently subjugated by Cromwell. Scotland is omitted, as it was only to be brought under the Commonwealth's control in 1654, incorporated into a unified Commonwealth of Great Britain and Ireland (also known as First British Republic).

The map is a copy of Christopher Saxton's *Anglia* with the addition of Ireland, and is the first major use of the map of the country as an image of the state itself. The idea of the nation state, defined by its borders and territory rather than its monarchy or ruling dynasty, is revealed as the revolutionary new political goal on the 1651 Great Seal through this image of Saxton's map.

mounting, and bound into books in series. Throughout the sixteenth century, printed maps were produced of both Britain (derived from manuscript sources and navigational information) and England, based on the first full survey of the country. This was the work of Christopher Saxton (a surveyor employed at somewhat second- or even third-hand by the English Crown) and who produced the first of his maps of the English and Welsh counties in 1574, maps he combined in 1579 into a full image of 'Anglia', included at the front of his atlas.

Saxton's maps and his atlas became the ultimate source of an immense publishing industry that provided maps of England and Wales cheaply and accessibly for several generations. By contemporary standards, they were not very useful (they did not include roads or topographical features in any detail) but what they did provide was an accurate image of an individual's place in the world, a sense of hierarchy that related their village or town to nearby cities, to the next level of governance after the parish or city authority, at county level and up to the national level based around the Queen herself in London or on progress across the country. From such images, available even to those who only had a rudimentary grasp of reading, and the widespread understanding of belonging to a greater whole, the modern nation state was born, based on the idea of shared citizenship of a place, and not simply subservience to a series of feudal masters.

Francis Bacon, the Elizabethan statesman and philosopher, expressed his belief that printing had changed "the appearance and state of the world".[2] As part of the print revolution, the general availability of maps—at least to the educated population—made possible a new way of imagining the specific place that each person inhabited, claimed citizenship of and (in part) owned. England and the English became, albeit briefly, aligned.

THE LANDSCAPE OF THE IMAGINATION

Hundreds of books have been written trying to define the nature of the English and Englishness. One of the most clearly identified and dominant themes emerging from that enquiry has been the idea of the English landscape, both as an archetype but also as possessing regional variety contributing to the whole, and its role in establishing a sense of belonging to a place that individuals didn't necessarily need to visit or experience themselves in order to feel part of. It became a fundamental aspect of the shared national idea of the country—an aspect that has led to a precise topographical literature and to landscape art being at the core of English painting.

If writers such as John Clare, Charlotte Brontë or Thomas Hardy are viewed as English, while being forever associated with a particular territory and terrain, and if artists like Gainsborough, Constable or Turner paint a particular place yet create an indelible interpretation of the English landscape in general, there must be a familiarity with the idea of the topography of England that to an extent encompasses and includes all its separate versions.

Anglia, 1579 Christopher Saxton

Courtesy the British Library/

Ashley Baynton Williams

After over 300 years without a new map of England,
based on on-the-ground measurement, Christopher
Saxton's collation of his individual county surveys
into a map of the whole of England and Wales was an
astonishing moment in mapping and emblematic of
Elizabethan confidence in its home realm. Saxton was a
little known Yorkshire surveyor when he started on the
task set out by William Cecil, Lord Burghley, Elizabeth's
secretary of state, on a commission from the courtier
Thomas Seckford, who was given a ten year patent on
the project. Astonishingly, Saxton completed the task
within five years, apparently working from hill peaks and
high buildings. Burghley wanted the mapping carried out
for reasons of practical politics and for ensuring the
defence of the country. His bound and hand-annotated
set of proofs of Saxton's maps is now one of the most
significant of the British Library's holdings.

The publication of this map, engraved by Augustine
Ryther, marks a point when the science of surveying, the
art of engraving, the politics and economics of the Tudor
period and the brief existence of the Kingdom of England
and Wales all come together and are triumphantly
defined. The map was widely copied and referenced for
centuries after and is the source for the map on the
Great Seal of 1651.

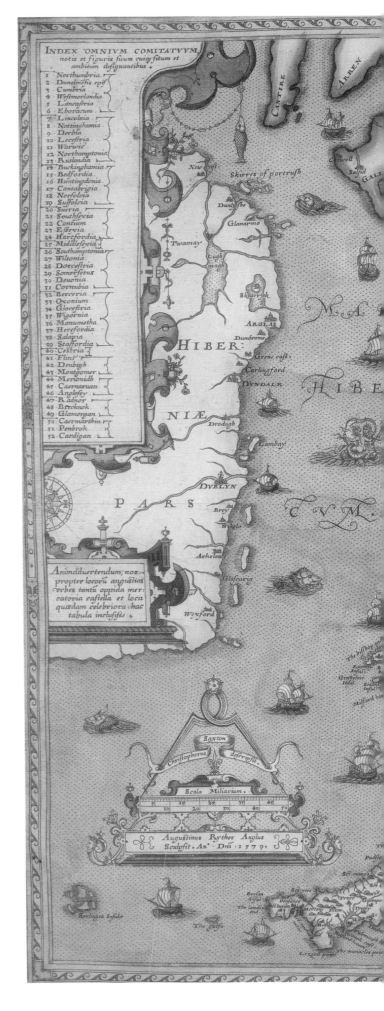

Despite the possibility of travel across the length and breadth of the country, relatively few people have ever comprehensively visited its tens of thousands of individual towns and villages, or seen its multiplicity of landscapes. The only means by which it was possible to imagine all these different aspects of England being one and the same territory was through the medium of maps.

Although the celebrated image of England that Shakespeare gives to John of Gaunt in his play *Richard II*, of a "precious stone set in the silver sea", would have been unintelligible to an earlier audience without an internalised and anachronistic map-like image of the island, any such image derived from the publishing efforts of the Tudor period (the great period of English landscape art and topographical writing) was post both the agricultural and industrial revolutions. The land had been thoroughly measured and assessed by the new breed of land surveyor and roads, canals and railways had been extensively cut through the countryside linking the whole country together. England was now a connected and intelligible commodity. The confidence of the landscape artist is born of the belief that the land was recordable and if necessary it could be conquered and 'improved'. If their audience did not need to visit the places being described, it was important for them to know that they could do so if the notion took them, whether by coach or train, and often within hours.

It was a confidence based on the detailed work of the early surveyors, the Military Board of Ordnance who had triangulated the whole country and who had started issuing the county map series from the start of the nineteenth century. The many early civil engineers also worked out the topography and geology of England, enabling them to dig mines, construct complex canal systems and build a comprehensive railway network. It was a confidence based on the manifest wealth-creating capacity of the surveyors' and improvers' efforts; it was also based on the unquenchable self-belief of the Victorians that they could dominate nature if they so chose—even if it meant turning those parts of the country and countryside most heavily industrialised into a burning, smoking, hell on earth.

Behind the landed gentry in Gainsborough's paintings is a re-invented and hard-working landscape, the source of their wealth. Constable painted the canal locks around Dedham in Suffolk that allowed trade with London and countries overseas, while Turner's landscapes celebrate the extraordinary impact of both human and machine on nature. They are descriptions of nature transformed, of a landscape after the surveyor has redrawn the map, has replanned the fields, manufactured new vistas, snaked canals along contour lines and hatched and dotted-in cuttings and tunnels. The English landscape as celebrated in the eighteenth and nineteenth century English novel (and by painters from the same period) is one that has already been captured and tamed. The flight of the Lakeland poets and the Pre-Raphaelite Brethren to the Lake District in reaction to the works of post-industrial society only highlights their difficulty in finding anywhere that was still as nature had left it. This, of course, ignores the role of the lake as training ground par-excellence for the surveyors working for the Ordnance Survey, testing all their capacity and ingenuity for accurate and consistent topographical recording.

MODERN ENGLAND

The twentieth century was not good for England. 200 years of being British had taken the wind out of the sails of the separate identity of Englishness. In a century of decline—of empire, global political leadership and power, economic strength and cultural confidence—huge efforts were put into the rescue of the British project, but the idea of England was allowed to become the receptacle for extreme nostalgia and sentimentality, becoming the eventual domain of the far right. London managed to maintain a different, more outward-focused trajectory, in which it was joined by several other large English cities by the end of the century. The country as a whole, however, appeared more interested in celebrating past glories; of merrie England, plucky little England, the heritage England, of busbies and teashops promoted by the then English Tourist Agency.

Cartography's main job during the twentieth century was to chart and to measure; to develop new ways of communicating information about the country and to reveal the statistical patterns of social and political issues, including employment, industry, agriculture, poverty, pollution and housing prices. It was also used extensively to investigate the past, to find and examine iron-age settlements, patterns of feudal agriculture and abandoned plague villages. In general terms, maps of Britain were for looking forward—for showing the new motorway network and major infrastructure projects such as the National Grid, representing Harold Wilson's "white heat of technology", or for representing the country's future place in Europe. Maps of England were for looking back—to the pageant of England, to former sporting glory and to find historic and nostalgic visitor attractions. Twentieth century maps developed to a high degree of accuracy and representation and by the end of the century had embraced the new digital age, but those of England had yet to explore this new territory.

However, just as the other national elements of Britain have begun to assert themselves and achieve degrees of independence with the start of the twenty-first century, so England is beginning to re-examine its identity. Its flag, the Cross of St George, has been largely retrieved from the xenophobic far right and English culture and identity is being reviewed in a new light. The multi-cultural, multi-ethnic map of England is being recognised as a strength, as seen in some of the map art produced by a wide range of artists or by maps celebrating the diversity that England can now offer. For instance, some of Layla Curtis' maps celebrate both England and Britain as microcosms of the greater world. There is every possibility that the mood maps of a future England will see it as an optimistic and forward-looking place that has great confidence in itself.

Typus Angliae, 1590, Jodocus Hondius

Courtesy the British Library/

Bridgeman Art Library

This highly decorative map depicts Elizabeth I at the top, covering a large portion of Scotland. She is flanked with Bible verses that extol her as the harbinger of prosperity. Below, an inscription dedicates the map to Robert, the Earl of Essex. Also included are symbols of art, war, industry and husbandry along with an English Nobleman and a Londoner in contemporary clothing flanked by their respective wives.

A highly political and personal image, Robert Devereaux (the Second Earl of Essex, who had had a brilliant and dashing early military career) had recently become a favourite of the Queen, clearly reigning supreme over the country below her. Essex's prowess as an artist, soldier, businessman and lover was widely extolled, but it is only with hindsight that we see his name attached to that of Ireland where he was to meet his military and political ruin only nine years later, devastatingly falling out of the Queen's favour and eventually losing his head a result.

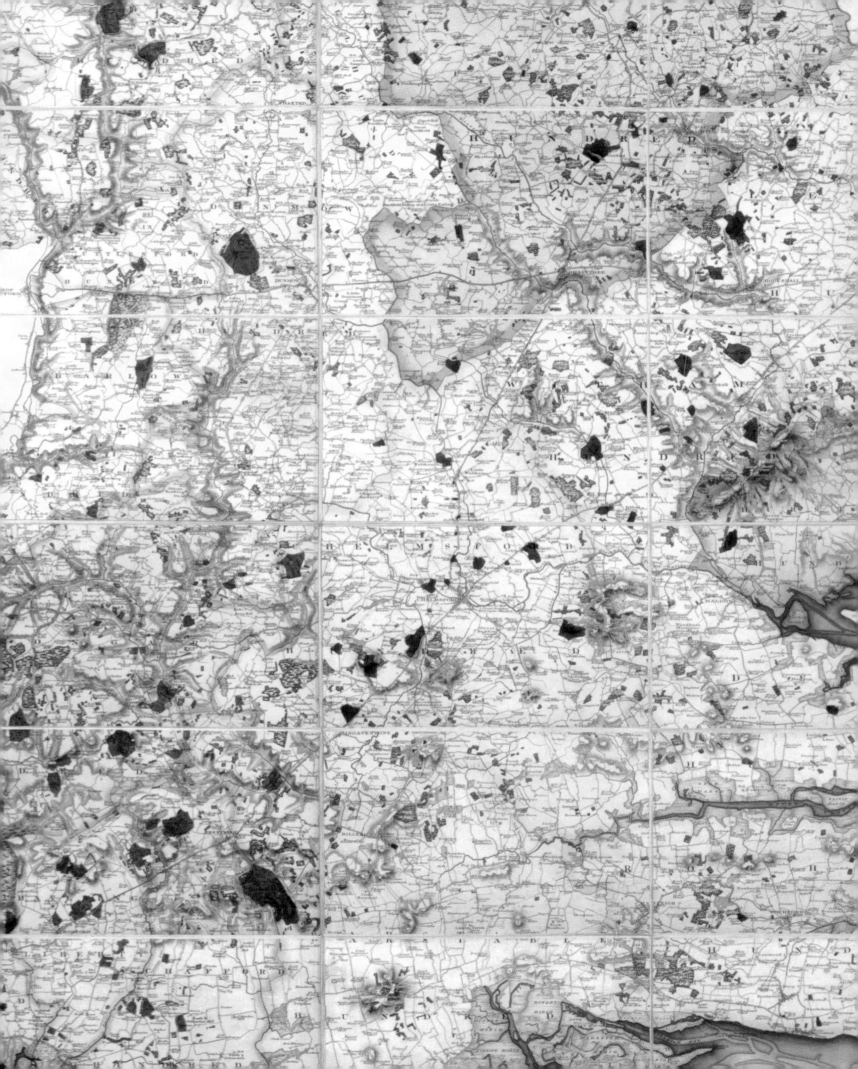

MAPPING
THE NATION

AT THE EDGE OF THE WORLD

THE ANGLO SAXON (COTTON) MAP OF THE WORLD

When the English author of the earliest-known map of the world composed his extraordinarily detailed view of the planet in the early eleventh century, he undoubtedly drew upon at least one Roman original as inspiration. Quite possibly, this was one commissioned in the first century by Marcus Vispanius Agrippa, the Emperor Augustus' most effective general and the builder of the original Pantheon in Rome.[1] More accurate and less metaphysical than later *mappae mundi,* it had a determinedly religious purpose and was bound into a book of *computus*—calculations for the precise dating of Easter each year.

In this map, the British Isles lie at the bottom left-hand corner to the northwest (east is situated at the top of the map as was the convention of the time). Their outline is instantly recognisable and the Orkney, Scillies and Channel Islands, with the individual Isles of Man and Wight are easily discernable. The British Isles adopt a position in the context of a wider world that stretches from the Taurus Mountains to the northeast—beyond which lions abound—and Africa to the south, essentially the Roman Empire. Britain (*Brittannia*) is clearly labelled, as are the cities of London, Canterbury and Winchester. In the exaggerated shape of the Cornish peninsula, two lone figures battle it out.

Even in the first millennium, England was a dominant political and societal concept. Saint Bede, from his Jarrow monastery in Northumberland, would celebrate the English as a 'people' in his *Historia Ecclesiatica Gentis Anglorum* of 731. Alfred the Great, 849–899, on drawing together the warring tribes that conquered the Roman *Brittannia*, would be one of the first to declare himself the King of such people. Following his seizure of Northumberland in 927, Alfred's grandson, Aethelstan, 895–939, was the first to convincingly make the same claim (in 928), although he also secured enough dominance over the neighbouring kingdoms of Scotland and Wales to grandly describe himself as *rex totius Britanniae* (King of all Britain).

MAPPAE MUNDI

The islands of Britain, Ireland and the Orkneys continue to be depicted as straightforward and simple entities on early *mappae mundi* and in other documents, such as the topographical works of Gerald of Wales, 1146–1223. However, by the mid-thirteenth century, maps would begin to refer to *Anglia*, *Scotia* and *Hibernia* as separate identities, with distinct political divides. In Matthew Paris' map of Britain, 1250, England and Scotland are connected by only the narrowest of bridges across a dividing sea, whereas in Richard of Haldingham's *mappae mundi* of 1300, the two countries are separated only by a short stub of a river that links the seas encroaching from both the east and west. Such was the nature of early Medieval maps, which—in seeking to express

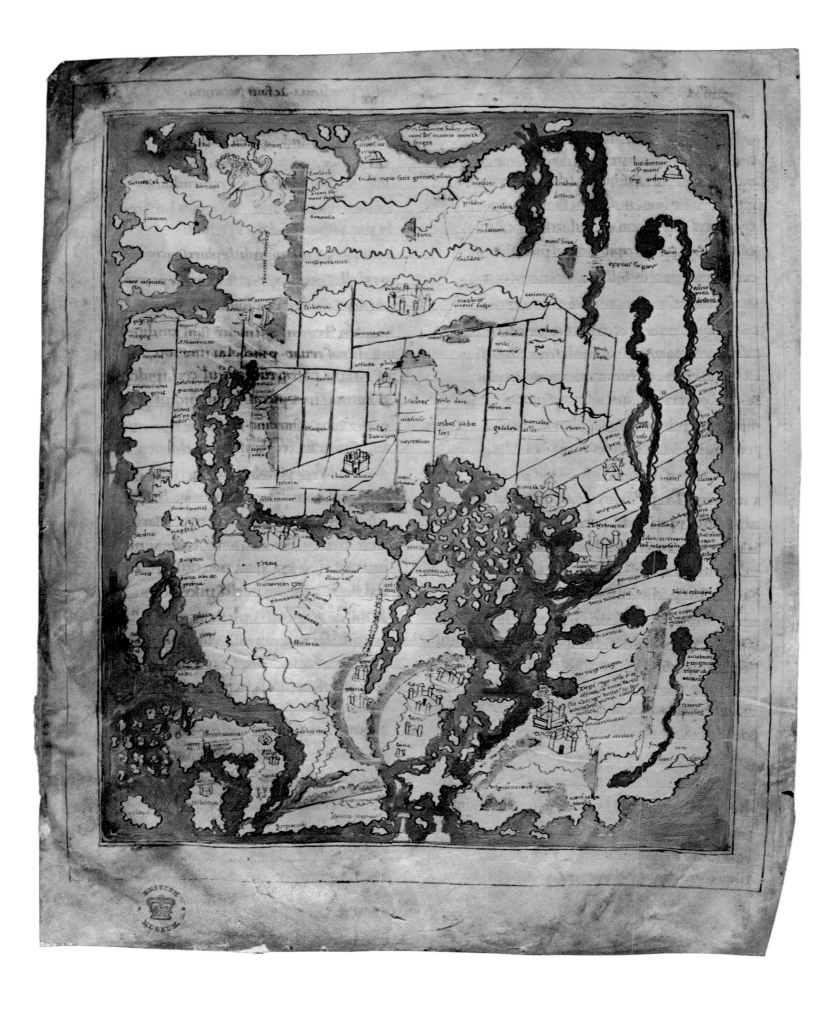

Anglo-Saxon *Mappa Mundi*, 1025–1050, Anon

Courtesy the British Library

Created in Canterbury between 1025 and 1050, this map is the oldest known surviving map of Britain. It probably derives from a Roman one recording the empire, of which Britannia was a province, and shows networks of lines that appear to be the boundaries of the imperial provinces. Also known as the Cotton Map (after the seventeenth century collector Sir Robert Cotton) it is noted for its complexity and considerable detail, including coastal areas, mountains, rivers, cities, towns and provinces. As with most other early maps, the Cotton is orientated with east at the top, but it also conveys secular origins in that it does not have Jerusalem at its centre, and by the absence of the Garden of Eden, which would have normally been included according to the early Medieval map tradition.

the political divisions within Britain as intense physical fractures—would invariably separate England and Scotland by turbulent and troubling waters.

MATTHEW PARIS

The title of Paris' earlier, less fragmented, map of Britain during the 1250s—*Britannia nunc dicta Anglia* (Britannia now called England)—refers to the change in titling of British countries from Roman terms to those acceptable under the new dispensation. Even then, the inherent triumphalism over the renaming of countries was not new; Paris was simply echoing the declaration of the tenth century chronicler, Aethelweard, that "Britain is now called England, thereby assuming the name of the victors".[2] In Paris' second map, he refers specifically to *Scocia: Ultramarina* (Scotland: Beyond the Sea); again in a pointed fashion.

The England of Paris' time was still adjusting to the Norman conquest of almost a century before. The nation that had emerged from the Anglo Saxon Heptarchy in the tenth and early eleventh centuries had precious little time to settle into existence and enjoy the fruits of its reasonably well-ordered administration, before Norman occupation forces would impose upon it. Despite its unwelcome presence, the Norman regime would effectively consolidate the new nation into a unified one, the facts and figures of which would be recorded in William I's great inventory of his new possession; *The Domesday Book* of 1086–1087.

If the Normans consumed England—with its system of law, coinage and taxation—and made it politically whole; they were also possibly the catalyst for a defined and widespread English culture as individual Anglo-Saxon factions coalesced in opposition to the common enemy and began to assert a single identity. In parallel, Norman culture, having fully and irrevocably transplanted itself to Britain, needed to adapt and come to terms with its new home. Over a period of two to three centuries, these two cultures would struggle and writhe in each other's grip, emerging as a distinctively English identity that was to effectively last until the reformation of Henry VIII, and the destruction of the English monastery system.

The architectural and artistic style of this period—the Perpendicular—is described by Nikolaus Pevsner, in his groundbreaking series of Reith Lectures of 1955 and subsequent book, *The Englishness of English Art*, as being "very much of England… and so much so that it lasted unchanged for nearly 200 years".[3] But such stability is inimical to the political imperative behind so much mapmaking. While England was still in ferment, it was necessary for Paris to link the country to the wider Christian world in his great pilgrim's itinerary (which he did by describing the journey graphically from London to Jerusalem), while

Greek Ptolemaic Map of the British Isles, circa 1300, Maximos Planudes

Courtesy the British Library/
Bridgeman Art Library

While working in Constaninople in approximately 1300, the Greek Byzantine monk and scholar, Maximos Planudes rediscovered a copy of one of the key books of antiquity, the 150 AD *Geographia* of Claudius Ptolemaeus (Ptolemy), a Roman/Greek scholar living in Alexandria. The original document had included maps, but as copyists subsequently had difficulty reproducing these only the first volume, a discussion of the data and Ptolemy's cartographic methodology, survived. From the description and the co-ordinates of the thousands of locations listed, Planudes re-drew Ptolemy's maps, including this one of the British Isles, with the characteristic Ptolemaic right-hand bend to the Scottish peninsula. Ptolemy would have used the academic resources of the ancient world to assemble his *Geographia*, including the celebrated library of his home city. If we don't have Ptolemy's maps, those recreated from his data are the next best thing, giving us a clear view of the knowledge available in the Roman world and above all, a picture of the ability of classical civilisations to visualise and accurately record the layout of the world in cartographic form.

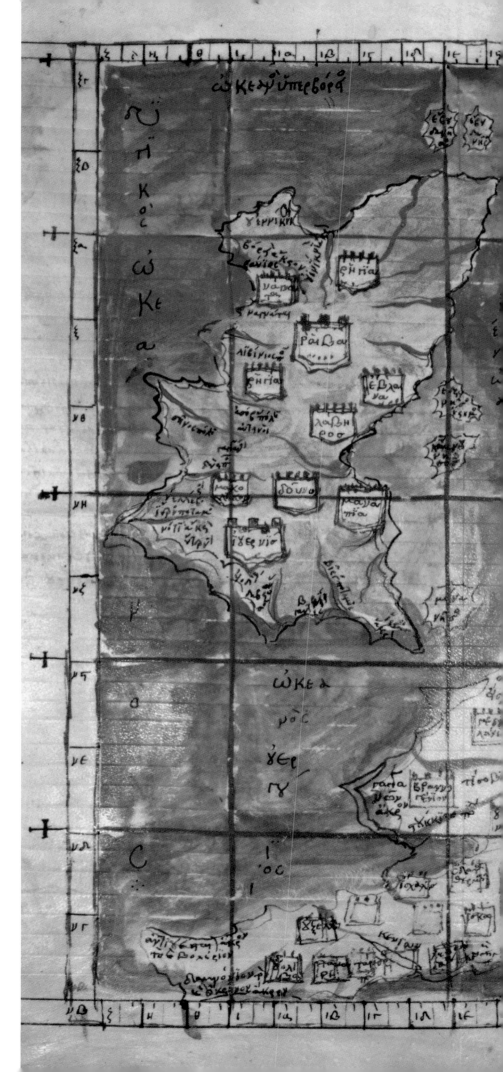

Hereford *Mappa Mundi*, circa 1290, Richard of Haldingham

Courtesy Hereford Cathedral/
Bridgeman Art Library

The *Mappa Mundi*—housed in Hereford Cathedral and signed by Richard of Haldingham and Lafford—is one of many examples of Christian world maps that visualise the known world in a circular form. There are over two dozen surviving English *mappae mundi* alone. Jerusalem is located at the centre, the eastern world (including the Garden of Eden) to the top and Europe and Africa to left and right, respectively, in the lower part of the image. Separating the continents is a T-shaped 'ocean' incorporating the Caspian and Mediterranean seas and known parts of the Atlantic Ocean. The characteristic resultant form is known as a T-O map.

The British Isles are shown in considerable detail to the bottom left of the map next to the outer, 'wild' edge of the world. Scotland and England are shown separated by a short stub of river with Ireland a separate island below, alongside depictions of a number of the other islands, including the Arran Islands and the Isle of Wight. Wales has its own identity and is separated from England by a central mountain with rivers flowing away from it northwards and southwards to the Irish Sea. Numerous cities (or Cathedrals/religious houses) are shown, including Hereford, Lincoln and York, illustrated by individual pictograms, as well as rivers and mountains. These reveal a high degree of geographic confidence—even within the distorting constraints of the *mappa mundi* form—and a clear understanding of the essential elements and disposition of England within both Britain and the world. The purpose of the map would have been to share this global understanding with as wide an audience as possible.

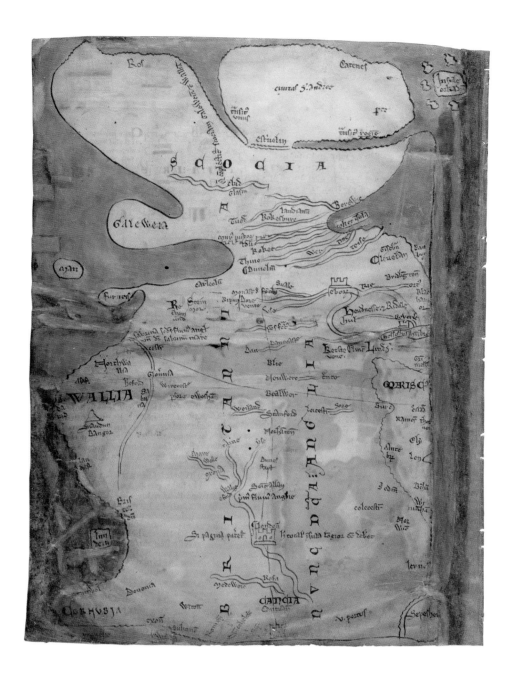

The British Isles except Ireland, in Matthew Paris' Map of the Itinerary from London to Jerusalem, 1250–1259, Matthew Paris

Courtesy the British Library

Matthew Paris was a monk and chronicler based at St Albans Abbey, just outside London. This map was included in a volume with his itineraries from London to Jerusalem and just as he gives over a page to the Holy Land at the far end of the journey, here he shows Britannia; emphasising the connectedness of one to the other. Paris' map includes England (*Anglia*), Scotland (*Scocia*) and Wales (*Wallia*), including the names of a number of counties

(such as Cornwall, *Cornubia*) and cities and towns (including Wymundham, a cell of St Albans). The Isle of Man and the Orkneys are also included. This is a definitive view of Britain as a whole, as a geographic place comparable to the surroundings of Jerusalem, although for a map in a book of roads he curiously chooses not to depict any of them.

Opposite **Map of the British Isles, late 1250s, Matthew Paris**

Courtesy the British Library

Paris' second map of Britain, this time intended to illustrate his *Historia Anglorum* (written around 1253), is a far more detailed work apparently constructed

around an 'itinerary' of towns (from Berwick to Canterbury) laid out—ageographically—as a vertical north-south axis down the map and forcing Suffolk into the location where we would expect to find Kent. His treatment of Scotland this time is far more circumspect, as not only does he include both Hadrian's and the Antonine Wall, described as walls (*murus*) dividing the Angols from the Picts and the Picts from the Scots, and reduces the connection with England to a narrow land-bridge, he also describes Scotland (*Scotia*) as being *ultramarina* (beyond the sea). This is clearly a very different, political, view of Britain in which England is the main player.

many *mappae mundi* would anchor the country in a religious setting that included the Garden of Eden. However, as the climate in England became more stable, Medieval mapmaking diminished, leaving only one other great flourish behind it, the Gough Map of 1360, a hugely influential map which is now housed in the Bodleian Library in Oxford.

THE GOUGH MAP, 1360

This map, named after Richard Gough who donated it to the Bodleian Library in 1809, is the only significant English map to have survived from the entire fourteenth century. It is by another unknown hand, but one that was exceptionally well-informed. In this map, England has a largely recognisable outline, whereas Scotland is entirely diagrammatic and Wales lacks the, now familiar, concave coast of Cardigan Bay. Inland, the rivers are well marked, as are the settlements, which are coded according to size and linked together by thin, red lines. The detail employed in the representation of land, in comparison to the relatively undeveloped coastline, is conclusive evidence that the map was compiled from land survey information and that the author probably had little or no access to the far more accurate Portolan charts that seafarers were using to navigate at this time. Internal navigation was clearly important—whether on precarious paths and roads inland or, more conveniently, by river—but coastal navigation was left to others.

Despite its significance, the Gough Map makes little distinction between the different nations of the British Isles. Political boundaries are noticeable by their absence—the only difference between England, Scotland and Wales being the modification in the level of detail and of accuracy. This may be an indication that the origins of the Gough Map are much older than its manufacture (which dates between the construction of the exterior walls of Coventry in 1355, and the renaming of Sheppey to Queenborough after Queen Philippa, wife of Edward III in 1366). The historian, Daniel Birkholz, speculates that the map was conceived in response to Edward I's violent annexation of parts of Wales and Scotland and his attempt to make Britain "a single monarchy". (Alternatively, it may reflect an older view of *Brittannia* and amount to a fourteenth century view of a Roman approach to cartography.)

Birkholz argues that the absence of borders and boundaries in the Gough Map is a marker of its geopolitical significance—representing an English Imperial approach to mapping the British Isles. It could also be perceived as a strictly geographical and topological tool, however this seems unlikely in the highly politicised climate in which it was made. It is this very uncertainty as to its purpose that continues to fascinate. Whatever its true significance, the very existence of such a large scale and highly accurate map appears to have satisfied the appetite for British Isles maps for many years to come—no further maps of significance were produced for almost another 200 years—at least as far as we know.

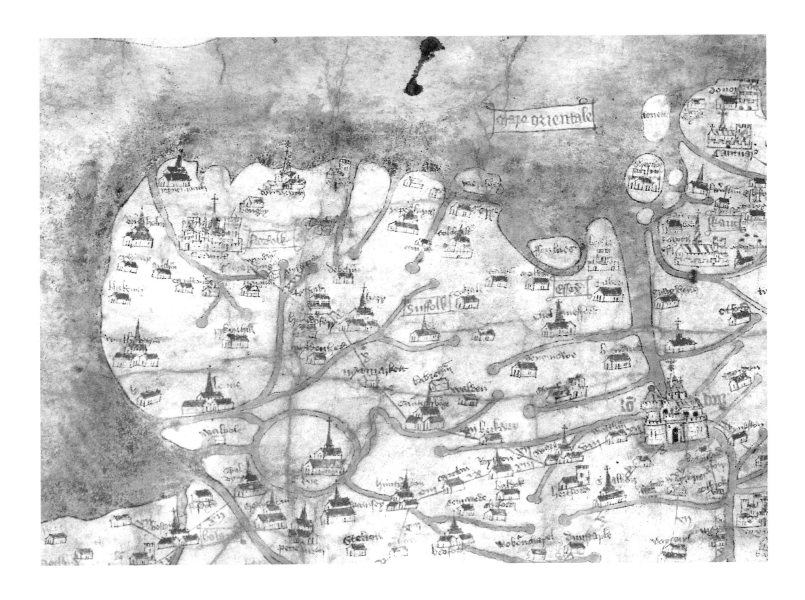

Above and overleaf **Map of Great Britain,
circa 1360, Richard Gough**
Courtesy the Bodleian Library
Often described as the oldest surviving road map of
Britain, the anonymous Gough Map could have been
created at any time between the construction of
Coventry's city wall in 1355 and the renaming of the
town of Sheppey in 1366. Its earliest known sighting is
when the map was bought by the antiquary Richard
Gough in 1774 for half a shilling from Thomas Price's
collection and then later donated by him to the
Bodleian Library in Oxford in 1809.

The map is drawn in pen, ink and coloured washes
on two skins of vellum. It manages to incorporate 600
settlements, 200 rivers and a rudimentary and partial
network of roads and routes between settlements
shown as thin red lines, with written distances (possibly
in furlongs), that cover over 4,731 kilometres in length.
Strangely, many key roads of the time, including well
known ones, such as the Fosse Way and Watling Street,
are not shown and some parts of the country—notably
the southeast—are poorly connected with the area
around Lincoln, being completely disconnected from
the greater network.

The map was clearly well-used and may
have been deployed by Edward I to plan military
expeditions in Wales and Scotland in his campaign
to gain complete authority over the whole of
Britain. It was also well-known enough to have been
the source of many other maps and remained the
main source of the more accurate ones—at least of
England—for many centuries. Scotland and Wales,
though forming major elements of the map, are
both sketchily outlined as rough blob shapes with
relatively little internal detail.

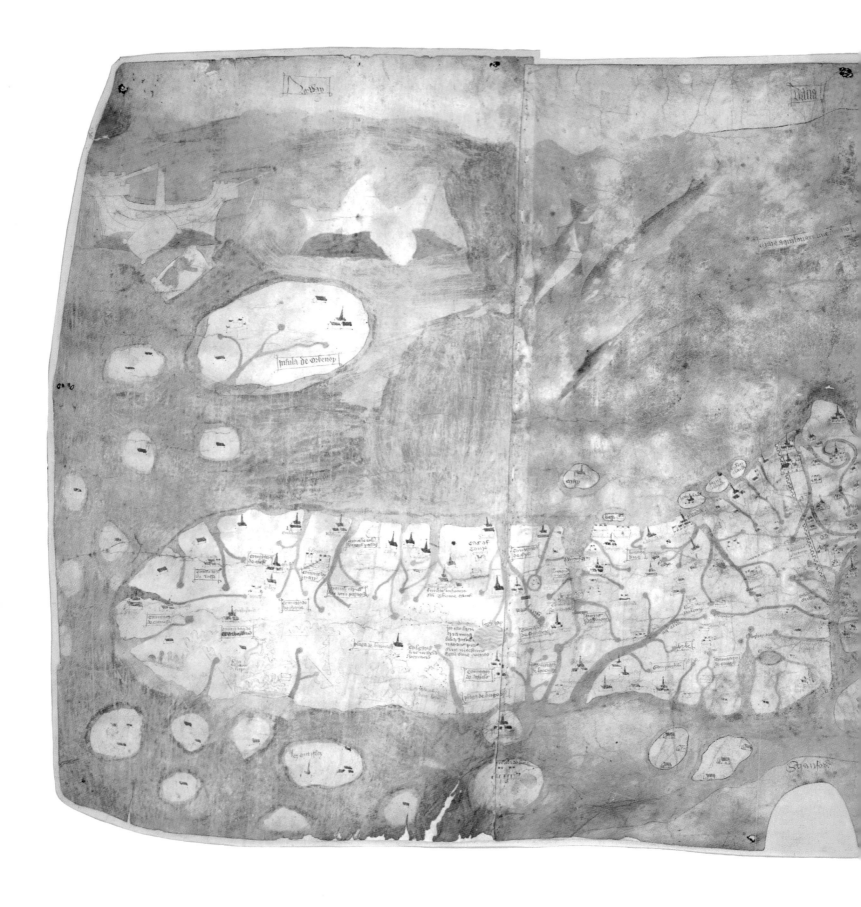

During this period the Gough Map would become a source of inspiration for copyists of new maps, both at home and overseas. In *The Gough Map: The Earliest Road Map of Great Britain*, 2007, Nick Millea suggests that Sebastian Munster's 1540 edition of the map (produced in Basel, Switzerland), Cladius Ptolemy's *Geographia*, George Lily's copperplate map *Britanniae Insula* (printed in Rome in 1546) and Gerard Mercator's 1564 map all derive from the Gough (although most of these managed to present an approximately correct shape for the coast of Scotland, if not Wales). There is also a sketch in a notebook belonging to an Essex merchant, Thomas Butler, 1547–1554, that is clearly directly based on the Gough.

The next survey of England was not to be undertaken until the 1560s and, by then, the reliance on the Gough Map for an accurate portrayal of the British Isles would have spanned a period almost impossible to imagine in our world of frenetic map use and production. During this time, England had 13 rulers—madmen, murderers and children between them—and had experienced one of the most violent periods in its internal history. It had lost (regained, and lost again) its French possessions, fought wars of succession (the Wars of the Roses, 1455–1487), suffered insurrection, invasion and battled against Scotland (Flodden Field, 1513) and across Europe (Harfleur, 1415, Agincourt, 1415, and Guinegate, 1478). The country was soon to crown Elizabeth I and would be entering a long period of unexpected stability. Mapping the nation was to become a major concern once again (in large part in order to ensure the country was secure from threat of further invasion). It is surprising that, even when the utility of mapping was familiar to the courtiers of Elizabeth I's court, in the period 1360–1570 there is little evidence of significant mapmaking when it would have been critical to the process of nation building, civil administration and, above all, to the art and practice of warfare.

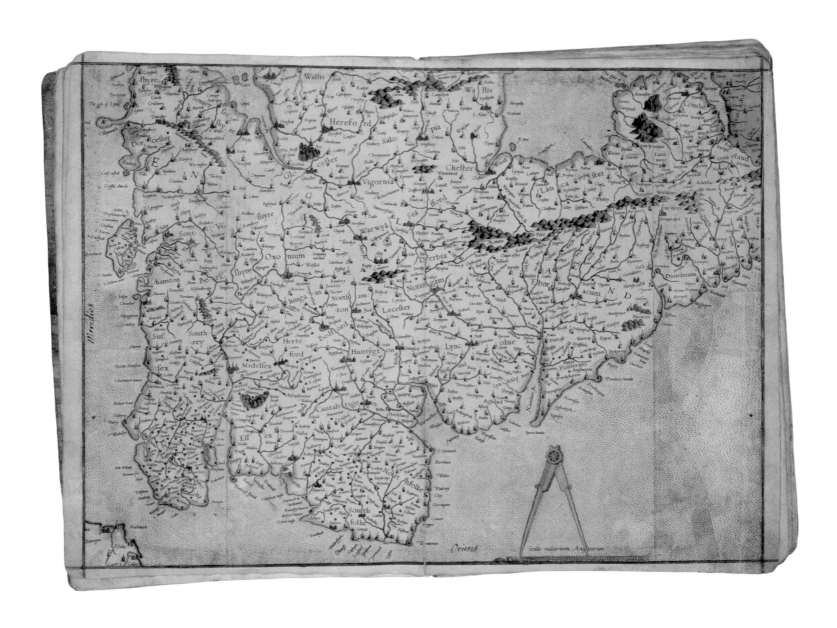

England from Mercator Atlas of Europe, 1564 (1570), Gerard Mercator

Courtesy the British Library

England from Mercator Atlas of Europe is a map from the atlas that Gerard Mercator compiled in the early 1570s. In it, the famous Flemish cartographer combined maps and images from a variety of sources while carefully removing any illustrations or unnecessary materials. The main source for this map is the work of Lawrence Nowell and John Rudd, master to Christopher Saxton, both of whom surveyed parts of England for Lord Burghley. Mercator's magpie-like working method is revealed in the discrepancies, including mislabeling (for example, listing Sussex as a village) and various key omissions, such as that of Windsor Castle.

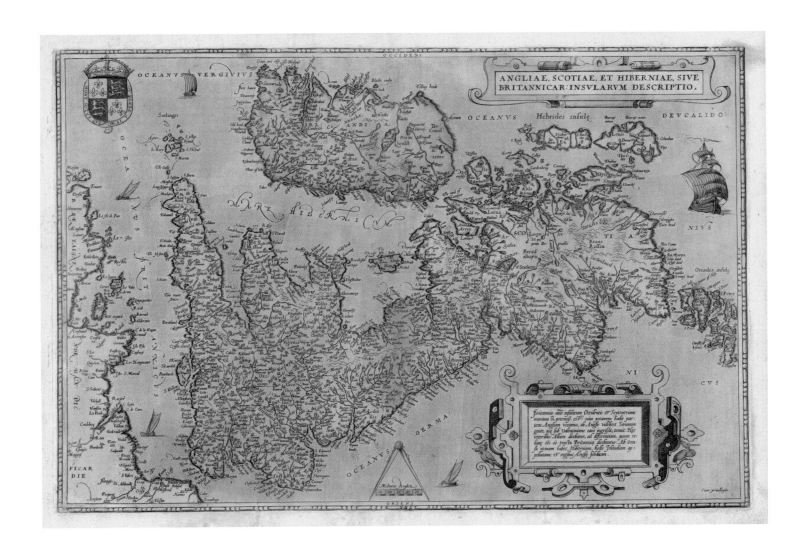

Angliae Regni Florentissimi Nova Descripto Auctore Humfredo Lhuyd Denbygiense, 1573, Abraham Ortelius

Humphrey Lhuyd created this engraved map of Britain with an elaborate cartouche, strap work, coat of arms and figures. This map, and other similar examples, were included in what is considered to be the first modern atlas (published by Abrahm Ortelius, a Belgium cartographer, in a supplement of the *Theatrum* in 1573). Ortelius borrowed from an array of sources, many of which are very rare, or no longer exist, and he helpfully included a bibliography identifying the cartographers whom he had borrowed from. Due to the multitude of source materials that Lhuyd conflated, there are some discrepancies in delineation and nomenclature in this rendering. Wales is divided into three traditional regions but its physical geography is almost double in width in some areas. Nonetheless, the atlas was in high demand until 1612 and succeeded in releasing 31 editions thereafter.

THE COUNTY SET

AGE OF EXPLORATION

As sixteenth century explorers traipsed the globe, cartographers would follow suit; surveying, measuring and, with luck, recording a return route which maps of the time had, so far, failed to offer. On his *mappae mundi* of 1507, Martin Waldseemuller updated Ptolemy's worldview to include the Atlantic Ocean and a coastline to the newly-discovered "America" (a title inspired by the explorer Amerigo Vespucci). Terrestrial globes (such as Martin Behaim's *erdapfel*, or earth apple, for example), first appeared at the end of the fifteenth century, however, such maps only really came into their own after the advancement of mapping techniques and the successful circumnavigation of the globe by Portugal's Ferdinand Magellan in 1522.

Suddenly maps were representative of the greatest advances in society; of science and intellectual curiosity; trade and commerce; adventure and derring-do; politics, religion and national pride; and, of course, colonial acquisition and sheer greed. Maps began to be used for decoration as painted wall hangings, tooled leather screens and woven tapestries. They became an integral part of the household interiors of the 'worldly-wise' who wanted to demonstrate this quality; symbols sufficient in themselves to indicate individuals of significance.

Such enthusiasm for maps would have remained a fashionable indulgence had it not been for the development of printing; first from woodblocks and later engraved copper plates. Suddenly maps began to appear in journals, pamphlets, books; Bibles even. William Caxton (the diplomat and writer who would be the first to introduce a printing press to England) printed a symbolic T-O representation of the world in the first scientific volume he published, the *Myrrour of the Worlde*, 1481, and a map of the Holy Land (quite possibly by Hans Holbein the Younger) was included in a 1535 English edition of the Bible. Most significantly, maps were printed on multiple sheets which, when assembled, could be mounted on rollers or hung on a wall—a decorative possibility for almost everyone. The best maps of the period were engraved and printed in Antwerp and elsewhere on the continent, until English craftsmanship began to catch up at the end of the sixteenth century. Once such technology and skill had taken root there was no holding back.

The technology of surveying had also begun to develop, with advanced geometrical theory arriving from the continent, particularly with the practice of triangulation. Triangulation made surveying more straightforward and diminished the requirement of a single baseline to measure ground distances. It did, however, require more sophisticated equipment for measuring precise angles and to accurately calculate north—something that many early and mobile compasses ere unable to do. Those surveyors able to master the mathematics and the equipment were clearly at an advantage over their more pedestrian colleagues, who were forced to laboriously pace out distances across the countryside.

COUNTIES

The cartographic revolution was far from the most significant event of the sixteenth century but it can be seen as emblematic of many of the great changes of the age. There was a spirit of confidence born of increasing stability, of social and scientific inquiry and of wealth and commerce, partly deriving from the unleashing of intellectual talent and of land and resources after Henry VIII's dissolution of the monasteries in 1536–1541. However, even as mapmakers drew and engraved their world visions, they were still reluctant to tackle their own country.

Individual landowners—due to practical motivations such as resolving boundary disputes and land management—were beginning to commission maps of their property. For centuries, title deeds were simply written descriptions—'terriers' and 'manorial extents', for example—but, by the beginning of the sixteenth century, a small proportion were supplemented with sketch plans and others with elaborate local maps (such as a plan of Sherborne in Dorset made for the Bishop of Salisbury between 1569 and 1578). It was not until the end of the century that mapmaking would become a commonplace activity for surveyors, a concept that was first recommended by land surveyor, Ralph Agas, in his treatise of 1596.[4] His proposal to an individual landowner to "know his own" was one adopted and echoed through innumerable later treatises and surveying manuals.

If landowners were slowly adapting to the idea of mapping their own property, the subject of mapping those areas owned by the Crown was an urgent issue for the government. During the reign of Henry VIII and Elizabeth I, there was a pressing need for maps of fortifications and of the vulnerable south coast but, following that first intensive foray into mapmaking, thoughts then turned to the rest of the country. This may, in part, have been due to the particular (almost obsessive) interest in maps shown by one of Elizabeth's advisers, William Cecil. Cecil was Secretary of State from 1550 to 1553, and 1558 to 1572, and later, as Lord Burghley, the Lord High Treasurer from 1572 until his death in 1598.

Cecil both collected and commissioned maps, and his extensive and well-annotated archive has survived largely intact. His interests cover the topics one might expect for the subtle government administrator and politician that he was—defence, taxation and communications—but he also appears to have appreciated his collection for its own sake; almost any map he encountered was bound into his own personal atlases. Cecil's interest in mapping would provide the impetus for recording the entire country cartographically (although his personal involvement in such a grand project is unclear). Initially, it was proposed that cleric, John Rudd, should spend two years travelling the country to perfect a "plat of England"—a project authorised by the Queen who instructed his employers, the cathedral chapter of Durham, to continue paying him during this period although the project and the map appear to have come to nothing.

In 1563, a second candidate was put forward for the job; one Lawrence Nowell, a member of Cecil's household who had produced a few small-scale maps of both England (later bound into one of Cecil's atlases) and Scotland. His involvement proved to be fruitless, however, quite possibly due to the lack of resources necessary to commission such a large scale project as well as Cecil's diminishing enthusiasm for the task. In the early 1570s, the concept would be resuscitated—a result of a private commission from the wealthy barrister, Thomas Seckford. This time, the surveyor was a one-time assistant of Rudd's: Christopher Saxton. Saxton was commissioned (with support and, ultimately, the reward of a lease of lands in Suffolk from the Queen) to prepare maps of the English counties and, later (from 1777–1778), the counties of Wales.

Very little is known of Saxton, either before he commenced this mammoth task or how he accomplished it. It is likely—if not certain—that he employed the new technique of triangulation, as "all Justices of peace mayours and others… within the several Shieres of Wales" were commanded "to see him conducted unto any towre Castle highe place or hill to view that countrey".[5] He may well have utilised the nationwide network of alarm beacons instituted in 1139 (and most famously lit to give early warning of the Spanish Armada in 1574) to achieve the task. Saxton's surveying work was complete in a total of five seasons—hardly a month per county—and would have necessarily been a rough and ready activity relying on what local knowledge he could glean about distances and his fairly rudimentary (by today's standards) equipment.

Saxton's maps—each printed from its copper plate almost as soon as it was prepared and proofs obtained by Cecil for his personal collection—are stylised, characteristic and relatively informative. Hills, rivers, villages and towns are the most prevalent elements with only an occasional stand of trees representing woodland, as well as a limited number of coastal features. Country houses are represented by their affiliated parks, but there are no roads, no key, grid or co-ordinates; all features that were to be found on European maps of the same time. The significance of Saxton's county maps, notwithstanding their attractiveness, lies in their topographical accuracy (with the exception of the shape and outline of Cornwall)—a result of his highly effective surveying work. When the 34 county sheets were published together in an atlas in 1579, it was possible to assemble and view a complete map of the country. In 1583, Saxton went one better, publishing a wall map of the entire country on just 20 sheets; presenting a more up-to-date, albeit less detailed, vision of England.

It is unlikely that the decision to map England county by county was Saxton's. Such an approach simply reflected the administrative *status quo* and the way the country was, and still is, divided and governed. It would, however, establish an approach to mapping England

Westmorlandiae et Cumberlandiae Comit

Courtesy the British Library

This map of Westmorland and Cumbria is from
the 1583 edition of the Saxton atlas of England
and Wales, first published as a whole in 1579.
Consisting of 35 coloured maps depicting the
counties of England and Wales, the atlas set a long-
lasting standard for cartographic representation
and the maps remained the basis for English county
mapping—with a few exceptions—until after 1750.
During the reign of Elizabeth I, the use of maps
became more common, with many government

matters referring to increasingly accurate maps
complete with consistent scales and symbols, made
possible by advances in surveying techniques.

Lord Burghley, (Elizabeth I's Secretary of State,
who had been determined to have England and
Wales mapped in detail from the 1550s) selected
the cartographer Christopher Saxton to produce a
detailed and consistent survey of the country. The
financier of the project was Thomas Seckford, Master
of Requests at the Court of Elizabeth I, whose arms
appear, along with the royal crest, on each map.

Overleaf **Four maps from John Speed's Pocket
Atlas of 1627**

John Speed re-issued his atlas of county maps,
Theatre of the Empire of Great Britain, in a pocket
version during 1627, reducing the counties to bite
size pieces for easy consumption by an eager public.
If the Tudor mapping revolution had made maps
familiar objects, Speed's final venture—he died
in 1629—made them fully accessible and readily
available to all.

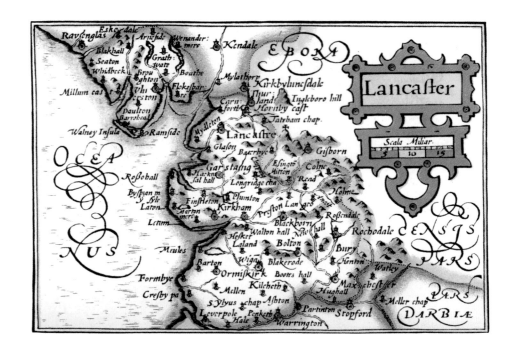

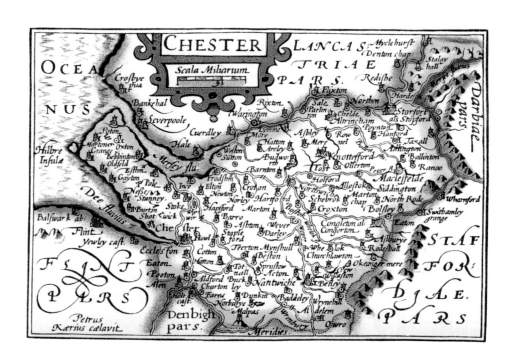

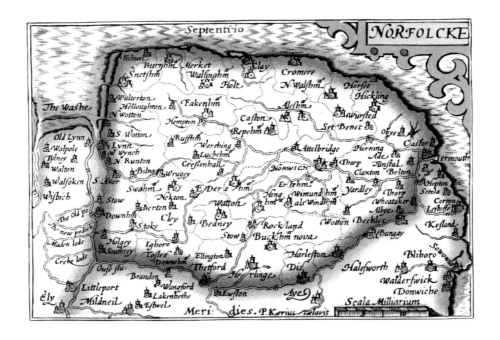

A map from John Speed's *Pocket Atlas* of 1627

Another example of John Speed's county maps of 1627. This map opened Speed's *Pocket Atlas* with a rendering of the entire country, with subsequent maps depicting specific counties in detail.

Overleaf **Map of the British Isles**

Angliae et Hiberniae accurata descriptio veteribus et recentioribus nominibus illustrata, 1605, Johannes Baptista Vrints

© Royal Geographical Society/ Bridgeman Art Library

A moment too late, Johannes Vrints has his map of England and Wales prepared when Elizabeth I dies and the throne passes to James VI of Scotland. To make the best out of his efforts, he elaborately dedicates it to the '*invincible James the Great, King of Great Britain, France, and Ireland*', and includes a genealogical tree to show the lineage from William the Conqueror to James I (of England).

that took another two centuries to shake off again: the vision of a fragmented country, where artificial county boundaries become the sole information content, a series of individual counties that terminate at their boundaries. Saxton also established the concept of the county print which, along with the hunting print, formed the sole decorative feature of any number of English homes, pubs and waiting rooms.

Saxton's county maps and his atlas of England and Wales (which brought them all together in a single volume) was only the first of many county mapping projects. John Norden, 1548–1625—again with Cecil's support—initiated his own surveys of the counties using greater, and more useful, detail than had been seen before, as well as a standardised key for different settlements and features. Like Saxton, he intended to publish maps of all English counties (to which he gave the general title *Speculum Britanniae*). Unfortunately he only succeeded in completing Hertfordshire and producing manuscripts of Cornwall, Essex, Hampshire, Surrey, Sussex and Northamptonshire—most of which were only engraved and published after his death. Norden's contribution to English mapmaking was in his use of the symbols noted in his key. He carefully differentiated between various ecclesiastical categories in his plotting of religious buildings and included details of heredity distinctions, building types and mineral deposits. His categories may have been idiosyncratic and personal, but the execution later fed directly into the practice of the Ordnance Survey and much modern mapping.

Not long after Norden's foray into mapping England, his friend, William Smith (following his return from the fertile mapmaking centre of Nuremberg), published maps of 12 counties from 1602–1603. There are obvious connections to be drawn between these maps and those made by Saxton, however, Smith's greatest influence in this respect would certainly be from Norden. Saxton's work had been groundbreaking because he had started from scratch; surveying England on the ground. His successors would apply themselves to the more tried and tested method of appropriation, and none was better at this than John Speed, whose atlas (Theatre of the Empire of Great Britain), was published in 1611 and included a full set of county maps. Speed's primary concern was to make a commercially successful product that could meet the public demand, while also concentrating on the finer details, such as aesthetic pleasure and factual corrections where necessary. His maps were largely based on Saxton's (being the only complete survey of England and Wales at the time), but were also augmented by surveys of Scotland and Ireland and were engraved by Jodocus Hondius in The Netherlands. Speed made no attempt to disguise this; on his map of Sussex, for example, he notes that the county has been "described by John Norden, augmented by John Speed" and elsewhere he similarly acknowledges William Smith.[6]

Saxton's maps remained the ultimate source of almost all regional and county mapping in England and Wales until approximately 1750.

The Bishoprick and City of Durham, 1610, from *Theatre of the Empire of Great Britain*, John Speed

Courtesy Bridgeman Art Library

This map shows the City of Durham and some of its hinterland. Highly detailed, it encompasses mountains and other such additions to provide a general overview of the lay of the land. Rivers are drawn out, and towns and villages around the city are clearly labelled. A detail of the castle grounds is presented in the upper right-hand corner along with its coat of arms and the surrounding river. The title cartouche, as well as the rose compass and the writing used, add to the overall decorativeness of the map. This work was hand-coloured and engraved by Jodocus Hondius.

The whole of the seventeenth century passed with hardly any new surveys (Jonas Moore's map of the southern Fenlands of 1658, John Oliver and John Sellar's survey and map of Hertfordshire of 1676 and Joel Gascoyne's nine-sheet Cornwall of 1599 being notable exceptions). In part, this was hardly surprising in a period beset by civil war and revolution, however, substantial advances were made in coastal mapping and in nationwide road mapping and the production of county maps proliferated in response to the public demand for cartographical artefacts of all kinds.

Saxton's maps also remained the basis for all government and administrative work during this long period and were used by politicians and civil servants at all levels. Saxton was even held responsible for inequities in taxation as George Owen of Pembrokeshire noted that he considered his county to be especially burdened by the Crown as a result of Saxton having given a single sheet over to Pembroke at a slightly larger scale (making it appear larger than neighbouring counties).[7] There remained official enthusiasm and royal encouragement for a new survey and a general understanding of its necessity, but—for a cash-strapped government that had been involved in many years of war—no resources were ever forthcoming.

ENGLAND REFORMED

Saxton published his great series at the apogee of England's burgeoning identity. The country had already absorbed Wales into its 'empire' under The Act of Union with Wales of 1536 although, taking into account the English Reformation of 1532 just a few years earlier, nation building and self-definition would have been in full swing for more than 30 years. Such elaborate construction, or the "great myth" as Edwin Jones would call it, would later unravel with Elizabeth I's death in 1603 and with the arrival of a new Scottish King, James VI, on the English throne. At least, during this period, the country would fulfil the promise of those words penned by Shakespeare's in *Richard II*, circa 1595:

> This royal throne of kings, this scepter'd isle,
> This earth of majesty, this seat of Mars,
> This other Eden, demi-paradise,
> This fortress built by Nature for herself
> Against infection and the hand of war,
> This happy breed of men, this little world,
> This precious stone set in the silver sea,
> Which serves it in the office of a wall,
> Or as a moat defensive to a house,
> Against the envy of less happier lands,
> This blessed plot, this earth, this realm, this England.[8]

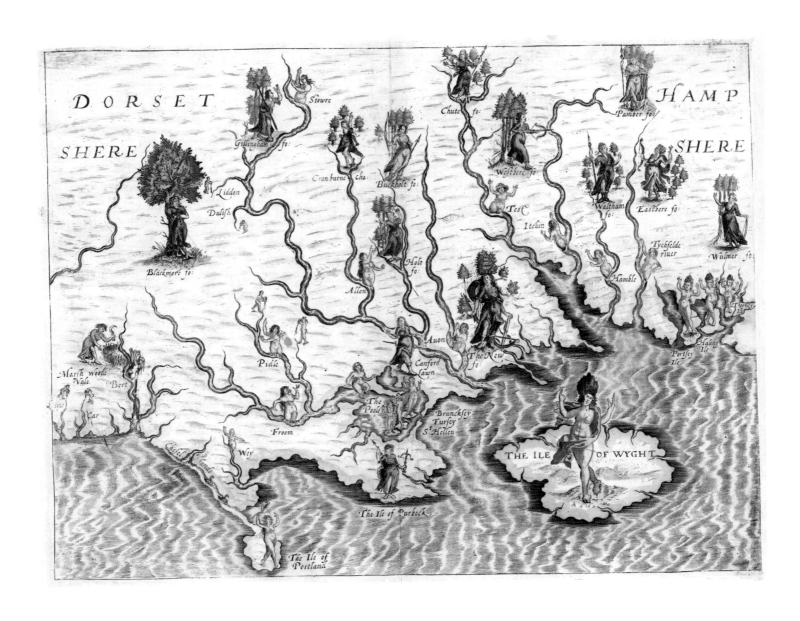

"Poly-Olbion", 1622, William Hole

Michael Drayton was a prolific Renaissance poet who produced the "Poly-Olbion" in 1612, a poem made from 30 songs. It was then reprinted in 1622 with several additions. Each song describes between one and three counties. The copies were illustrated with maps of each county by William Hole; there were 18 maps initially, followed by another 12 when it was reprinted. This map was designed as a visual journey around the kingdom, county by county. The description of topological features for England and Wales is pictorial and unrestrained by accurate representation. The main features are the nymphs and deities populated along the rivers and tributaries; and the towns and cities are symbolised by figures with castles and spires on their heads. Places are depicted anthropomorphically, meaning that the topological features are personified.

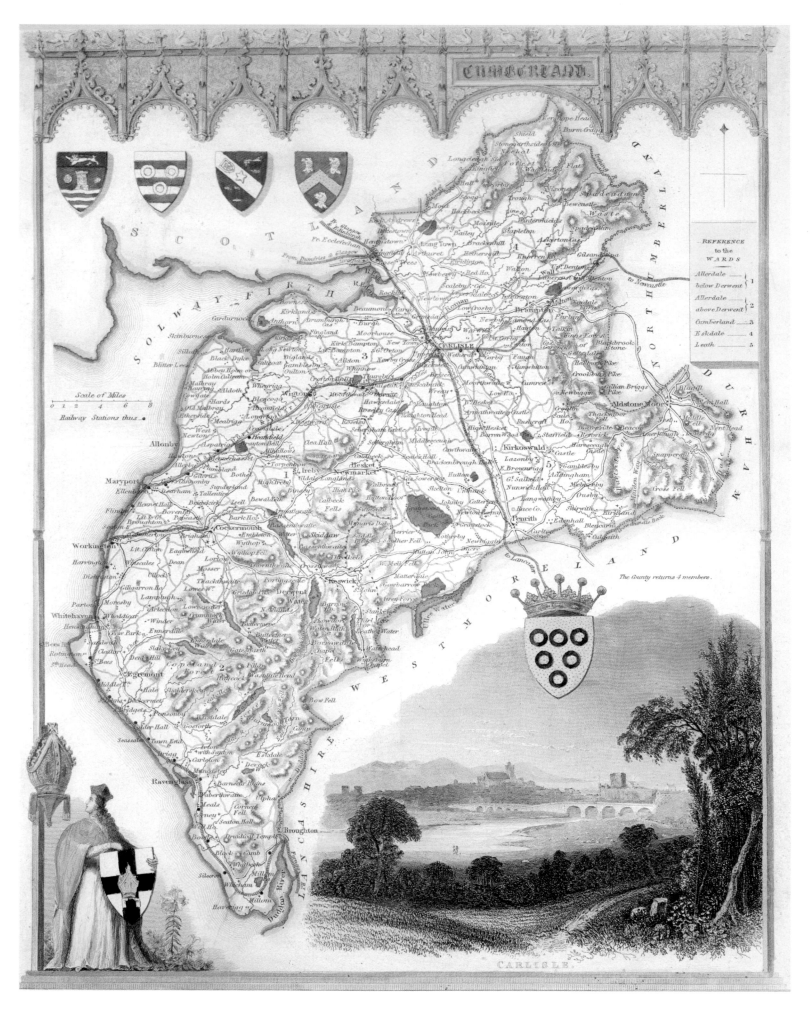

CUMBERLAND.

Map of Cumberland, from Moule's
The English Counties Delineated, **circa 1836,**
Thomas Moule
Courtesy Map House, London/
Bridgeman Art Library

This map was created by Thomas Moule, the scholar and writer on heraldry. Such interests are apparent in this map, as it includes several coats of arms along with images of noblemen and architecture. Moule's maps were renowned for their detail, making them highly sought after objects. He created this map, along with others, for each of the different counties in England. These were published in his two volume work entitled *The English Counties Delineated* and were fondly nicknamed the "Moules".

All the elements were in place for a new England: a country under threat from its neighbours; a king determined to establish a new national religion in defiance of the old, established and continental order; general wealth and prosperity; and the prospects of overseas trade and conquests to bolster the new order. If this caused internal strife then the extirpation of anyone who didn't share the vision for the new order was enabled— the propaganda machine went into full swing.

Saxton and his successors, Tudor and Stuart, were very much part of the new national determinism and identity forming. For the first time, with the loss of Calais in 1588, England had a well-defined physical shape that (apart from minor adjustments to the border with Scotland) has not meaningfully changed. Accurately measuring and representing England, and distributing such mapping worldwide, became a political act of huge significance; one that would produce a vision of the country that embodied and symbolised the nation as separate from the reigning monarch.

The image of England, as initiated by Burghley, surveyed by Saxton and then re-interpreted by Speed, is of a rural idyll of many small, interconnected towns and villages. London barely registers; the country is Shakespeare's demi-paradise writ large, with its pastel fringed counties and prosperous and settled appearance. It is a land about to be redefined by the Civil War and, as a result, cartographically presented from a very different perspective.

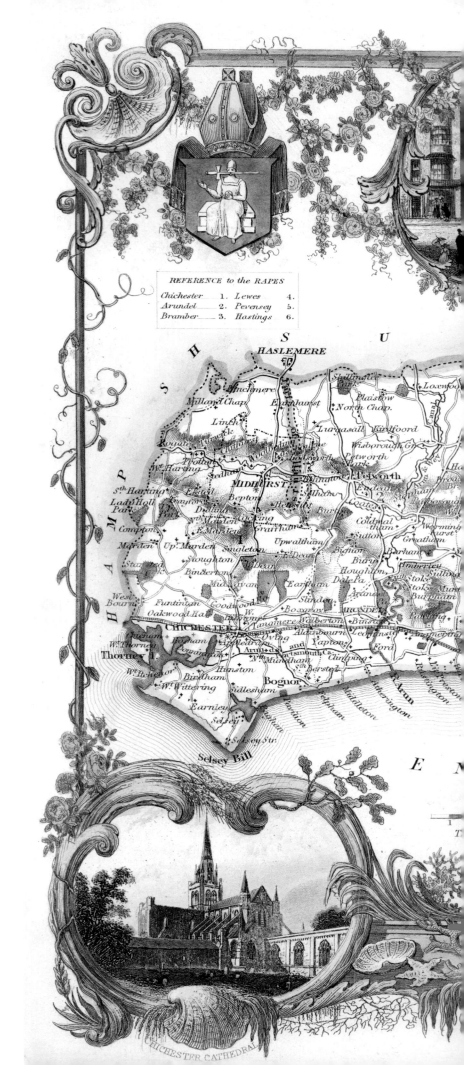

Map of Sussex, from Moule's *The English Counties Delineated*, 1836, Thomas Moule

Courtesy Bridgeman Art Library

Moule's maps were some of the last such decorative maps to be published. The county maps as a whole were first published in the 1830s during the reign of William IV. They were later reissued in 1841 with additions made to the original plates, such as the inclusion of railway construction. The hand-coloured steel engraved plates were then re-used in a James Barclay publication, titled *Barclay's English Dictionary*. Moule was a publisher of important works on heraldry and antiquities. In this map, we are able to see Brighton in panoramic view, serving as an indication of its popularity as a bathing resort at the time. Chischester Cathedral and Arundel Castle decorate bottom left and right corners respectively.

CHAIN PIER BRIGHTON

CHICHESTER

SUSSEX

ARUNDEL CASTLE

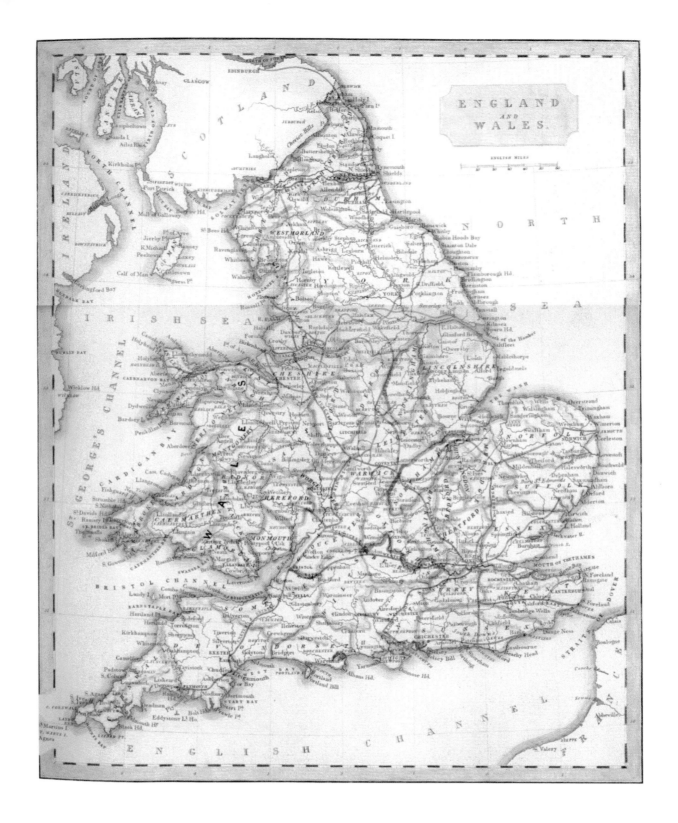

England and Wales, from Moule's *The English Counties Delineated*, 1836, Thomas Moule

This map was another published in Moule's *The English Counties Delineated*. Such maps were highly regarded and considered the most beautiful of their period, showing towns, villages, parks, hill, rivers and canals in all their intricacy.

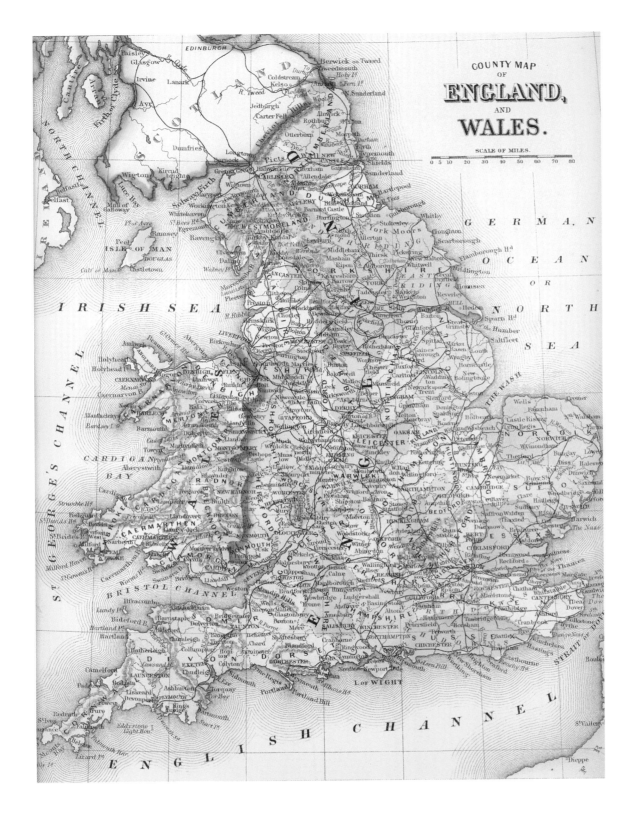

County Map of England and Wales, 1890,

Samuel A Mitchell

From the most popular of American nineteenth

century atlases, Samuel A Mitchell's *New General Atlas*,

was published in Philadelphia from 1866 onwards in

numerous editions.

**A map of the county of Devon, 1765,
Benjamin Donn**

Courtesy Ashley Baynton Williams

Benjamin Donn's map of Devon was the first recipient of the Society of Arts premium for a county map at one-inch to one mile; a prize set up in part to counter advances in surveying and mapmaking in France. Donn's 12 sheet map of Devon set new standards for accuracy, using trigonometry to achieve a greater level of detail. Five categories of roads are shown, along with details of settlements (including two town plans), with even remote farms and cottages depicted. This section of Donn's map shows the Membury area.

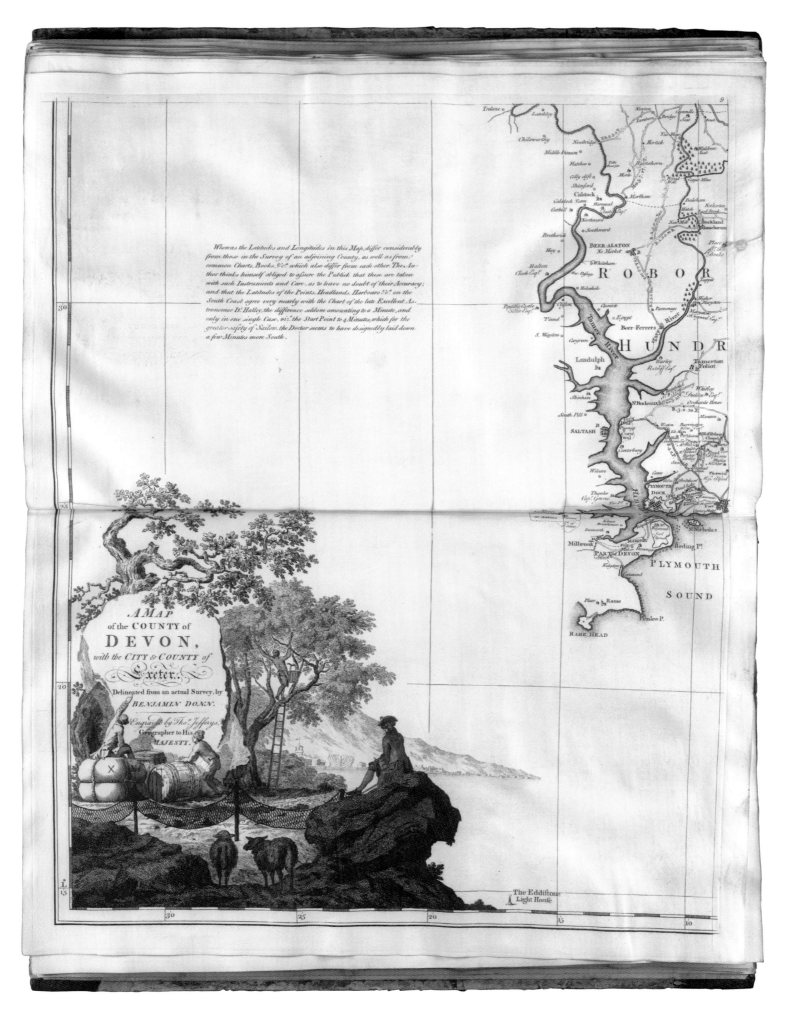

Whereas the Latitudes and Longitudes in this Map, differ considerably
from those in the Survey of an adjoining County, as well as from
common Charts, Books, &c.ᵗ which also differ from each other. The Au-
ther thinks himself obliged to assure the Publick, that these are taken
with such Instruments and Care, as to leave no doubt of their Accuracy;
and that the Latitudes of the Points, Headlands, Harbours &c.ᵗ on the
South Coast agree very nearly with the Chart of the late Excellent Astronomer Dr Halley, the difference seldom amounting to a Minute, and
only in one single Case, viz.ᵗ the Start Point to 4 Minutes, which for the
greater safety of Sailors, the Doctor seems to have designedly laid down
a few Minutes more South.

A MAP
of the COUNTY of
DEVON,
with the CITY & COUNTY of
Exeter.
Delineated from an actual Survey, by
BENJAMIN DONN.
Engraved by Thoˢ Jefferys,
Geographer to His
MAJESTY.

The Eddiſtone
Light House

The County of Essex, 1824,
Greenwood, Pringle & Co

Courtesy Ashley Baynton Williams

John and Christopher Greenwood, started their
county mapmaking business in 1817, and it initially
thrived in the hay-day of commercial mapmaking
until it finally succumbed to the overwhelming
influence of the Ordnance Survey during the 1840s.
In the1820s, they were busy producing a series of
high quality maps including over 12 different counties
between the years 1822 and 1826, of which this
depiction of Essex is one. Despite the competition,
the Greenwoods benefited considerably from the
triangulation work that the Ordnance Survey had
done across England (and which it allowed private
mapmakers access to) allowing them to effectively
compete in the new mapmaking market.

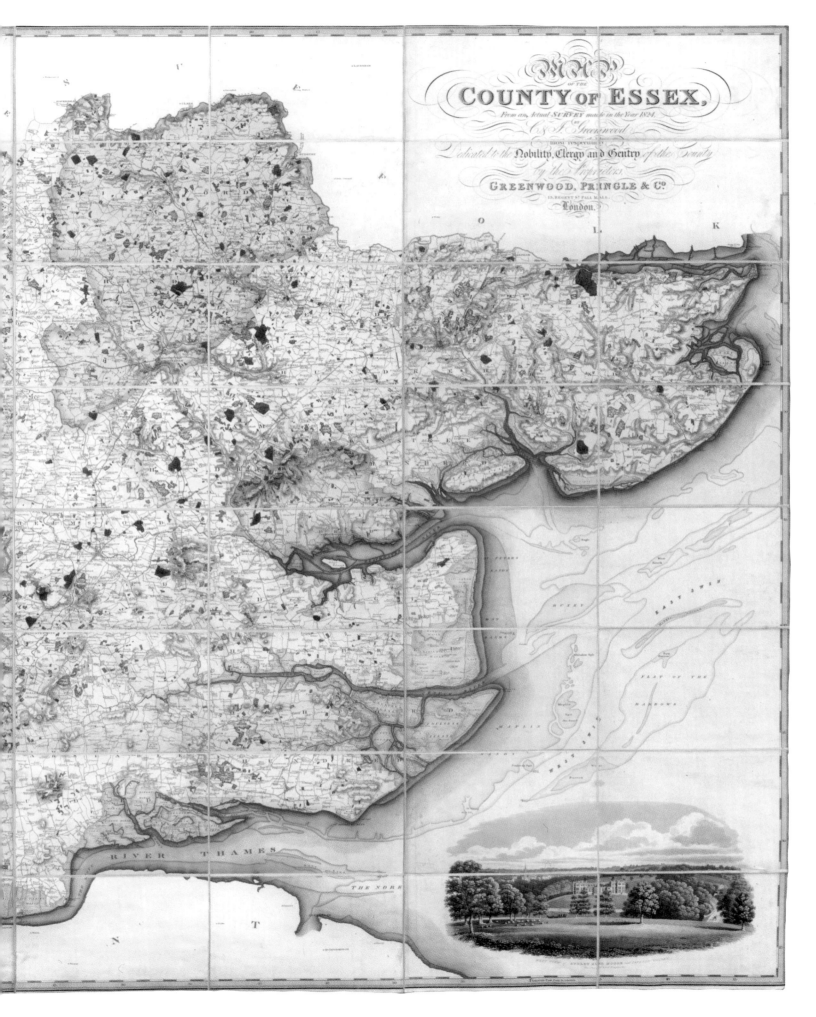

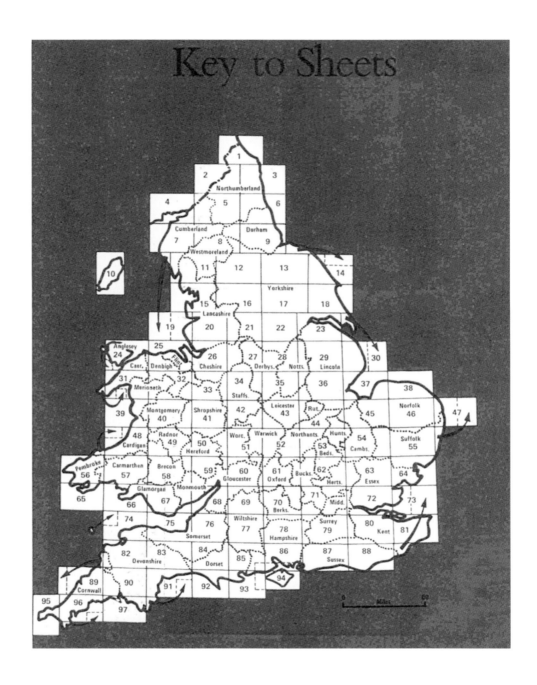

**Back cover of the First Edition of Ordnance Survey
one-inch to one mile map series 1805 to 1873**

THE ORDNANCE SURVEY

The English Enlightenment brought in a new approach to precision and measurement that had a major impact on surveying and mapmaking; although it was by no means a rapid transition. Old habits among county surveyors died hard, and ideas that arrived from the continent, in part with the new regime of William of Orange after the Glorious Revolution of 1688, spread only slowly out from London. The county remained the standard topographical unit and each of these was the fiefdom of different surveyors, keen to prevent interlopers from intruding. New ideas and techniques were picked up by enthusiastic, less experienced, surveyors, but possibly only deployed when they were freed from a long apprenticeship to set up their own concern.

Jonas Moore, an early member of the Royal Society in London (and therefore party to the intense discussions between the leading astronomers, mathematicians and scientists of the day) led the way in his survey of the southern part of the fens. Commissioned in 1650 by the Adventurers' Company, the survey was a very practical venture intended to bring the highly fertile fenland into agricultural use. Fortunately for posterity, the company also wanted publicity and advised that "when there is a perfect Mappe made… to have it printed".[9] Moore's Mapp of the Great Levell of the Fens was drawn at a scale of two inches to the mile—an act of astonishing confidence in his own capacity—and was published by 1658, complete with the Commonwealth coats of arms. The Restoration of the Monarchy in 1660 meant that his new map was

politically untenable but it remained the most accurate record of the area until the Ordnance Survey arrived there in the nineteenth century. Moore's own skills put him in greater demand under the new regime and he was rapidly commissioned by the Navy Board to map The Thames "from Westminster to the sea" and, in 1665, was appointed to be Assistant Surveyor of the Ordnance in the Ordnance Office. He took over as Surveyor General in 1669.

The Ordnance Office was charged with maintaining the supplies to the military but also the maintenance and construction of its buildings and engineering infrastructure. Its surveying expertise ultimately led to the establishment of the Ordnance Survey in the late eighteenth century under Major General William Roy. Moore's influence may well have been channeled through the Ordnance office (later the Board of Ordnance) to the Ordnance Survey, but his work on the Fenland survey left no other immediate influence. Many of the maps of this period, however, continued to be copies, an activity so widespread that one notable surveyor, Peter Burdett, admitted in 1768 that "it is true that new maps of England are daily published, but it is equally notorious that they only serve to transmit to us the errors of those from which they were copied, and generally with new ones."[10]

Numerous surveyors were at work in the first half of the eighteenth century; extending their regular vocations as estate surveyors to the far more ambitious goal of mapping whole counties. Necessarily, they relied on existing sources that they then supplemented by varying degrees of new fieldwork. Their standards of presentation were variable as were the scales to which they worked. Particular individuals stand out, such as Henry Beighton (who published his highly acclaimed map of Warwickshire in 1728 at the innovative—and inspirational—scale of one-inch to the mile) and, above all, John Rocque who published maps of Shropshire (in 1752), Middlesex (in 1754), Berkshire (in 1761) and Surrey (posthumously in 1768). With a background in landscape surveys for aristocratic clients and a self-proclaimed *dessinateur des jardins*, Rocque is best known for his elegant multi-sheet maps of London. He worked both from his original surveys but also published very stylish and stylised images concentrating on the—not entirely accurate but always compelling—texture and use of the land.

The work of Beighton and Rocque—compared with the paucity of others' efforts—may, in part, have inspired Henry Baker, a founder member of the newly (in 1754) established Society for the Encouragement of Arts, Manufactures and Commerce (better known as the Society of Arts and, later, the Royal Society of Arts), to insist that the Society took action. In both 1759 and 1762 it advertised that:

The Society proposes to give a sum not exceeding 100 pounds, as a gratuitity to any person or persons, who shall make an accurate survey of any county upon the scale of one inch to a mile; the sea coasts of all maritime counties to be correctly laid down together with the latitudes and longitudes.

13 prizes were awarded to mapmakers between 1765 and 1809 with the first going to Benjamin Donn for his 12-sheet map of Devon depicting, in exceptional detail, roads, mines and landscape information.

The following prize went, in 1767, to Peter Burdett. Based in Derby, Burdett was part of a circle of scientifically-minded friends that included the painter Joseph Wright. He was, in many ways, a true scion of the Enlightenment; widely interested in the greater world and engaged in scientific enquiry (he features in several of Wright's paintings of experiments). He was a methodical mapmaker, revealing a unique view of Derbyshire in the 1760s, only marginally changed at that time by enclosure. He was also careful to show his techniques from a diagram of his triangulation and compasses showing both true and magnetic north.

The Society of Art hoped to inspire the complete remapping of the country to a consistently high standard, however, by the end of the century and the publication of the first Ordnance Survey map in 1801, only 30 counties had been covered to their satisfaction. Several of those surveys then formed the basis for individual maps in the Ordnance Survey County Series. Despite this, the Society undoubtedly set a high bar for the Ordnance Survey to build on.

The 'Mudge Map' of Kent, 1801, William Mudge

This map of Kent, printed at the scale of one-inch to the mile in 1801, is recognised as being the first Ordnance Survey map, although published by William Faden rather than on the OS' behalf. Following the French Revolution, the growing fears over an invasion threat from across the channel lead to the English government commissioning a military survey of the vulnerable south coast using precise measuring equipment so that adequate defences could be planned. As a military map, depicting areas of the land that could be used as an advantage in battle was of importance, and is shown by the intricate three-dimensional hill shading, along with a clear and accurate depiction of communication routes such as rivers and roads. Its intricate depiction of the terrain and the technical ability of the surveyors can be seen to demonstrate Britain's political and geographical strength. Colonel William Mudge was the Superintendent of Ordnance Survey and oversaw the time consuming (work began on the project in 1795) and laborious use of measuring equipment used to produce this map. His later work on surveying Black Comb in Cumbria during 1807 and 1808 is thought to have provided inspiration for the poet William Wordsworth.

Ordnance Survey Map of Essex,

County series, circa 1870

A later version of the same area of Essex from the county series Ordnance Survey map. The topography is now much more clearly marked and several railway lines can clearly be seen running across the landscape. Despite the relatively late date, contour lines have yet to appear on this map.

The arrival of the Ordnance Survey did not immediately bring the private county mapmaking business to a halt (it would continue for approximately another 50 years), however, its nationwide trigonometrical data published in three volumes, from 1799 to 1811, brought new accuracy to their endeavours. In fact, the Ordnance Survey and such firms worked together, in large part to ensure that their activities did not conflict any more than necessary. The private companies, on the other hand, carried on with their well-rehearsed turf-wars and internecine battles. The best-known names among them included the brothers Christopher and John Greenwood, William Faden and John Cary. The latter two had won Royal Society awards, while the Greenwoods published their *Atlas of the Counties of England* (containing all 42 English counties), in 1834. The death of Christopher Greenwood in 1855 brought the era of competitive and semi-piratical mapmaking to an effective end.

The concerns of the eighteenth and early nineteenth century mapmakers had turned away from the nation-building of the Reformation period and were overwhelmingly driven by the dictates of commerce and the demands of a general public largely spoilt for choice. The countervailing influence of the English industrial enlightenment— the age of reason—and of the hunger for 'useful' knowledge managed to turn several of those projects to good effect but the results were perhaps not as inspiring as they might have been. English mapmaking

had achieved far more ambitious results elsewhere, especially in India, where the Great Trigonometrical Survey commenced in 1802 and, by 1856, had established the precise co-ordinates and height of Peak XV (later named Mount Everest after George Everest, second Superintendent of the Survey of India).

In England, however, the county system for map organisation was only slowly being replaced by the grid-based, one-inch to one mile Ordnance Survey maps (now known as the Old Series). The old certainties were still intact while the world was changing at a bewildering pace. When England entered the second half of the nineteenth century, and the high Victorian period, it was also embarking upon an age of management.

The origins of the Ordnance Survey lie in the Jacobite Uprising of 1745. William Roy, an officer in the quartermaster's office, was put in charge of surveying and mapping the highlands of Scotland—an exercise that resulted in the Duke of Cumberland's map. In 1763, having just returned from the Seven Years War, Roy proposed to follow his work in Scotland with a similar survey of England and Wales—an exercise that was dismissed by Parliament as "a work of much time and labour, and attended with great expense" and one that was hardly welcomed as it would have inevitably trespassed on the country estates and holdings of many MPs. Roy tried again with a second, more modest, proposal in 1766, but was turned down once again.

Roy was director of the Royal Engineers when the French laid down a challenge to national pride by accusing the English of being inaccurate concerning the exact latitude at the Greenwich Observatory in 1783. Roy reacted to such a challenge by seeking to accurately map the relative positions in both latitude and longitude between the Paris and Greenwich Observatories on behalf of The Royal Society. He proposed to set out a system of triangulation across the channel to connect the two observatories, starting with a five mile baseline that he painstakingly charted across Hounslow Heath, below the present Heathrow Airport, using a system of accurately dimensioned glass tubes. By 1787, he extended a system of triangles from his baseline down to the Kent coast, allowing the French engineers to connect them across to the Pas de Calais and on to Paris.

As Roy explained to The Royal Society in 1786, this exercise had a greater significance than the limited goal of comparing the positions of two observatories, "namely, the laying of the foundation of a general Survey of the British Islands". Within only five years, the threat from Napoleon had proved the urgent necessity of at least using the triangulation to survey the highly vulnerable south coast. Roy died in 1790 before this decision was made, but his legacy was assured. Two of his deputies, Major Edward Williams and Lieutenant William Mudge, began the survey of Kent for the Board of Ordnance in 1795, publishing their map in 1801.

Williams proved an uninspiring leader of the fledgling service (then called the Trigonometrical Survey) or "easily, the worst" in the words of his successor Charles Close Mudge. Mudge, who replaced Williams in 1798, was rather more successful in the role, leading the 'Great Triangulation' across Britain and Ireland; a job that involved building permanent trigonometrical stations across the (often challenging) terrain of the country.[11] Scaffolding was erected on the top of church spires, including the dome of St Paul's in London, for the same purpose. It was a project that continued until 1853, defining the skeleton of the map of Britain.

This was a map of Britain—not England—they were making, eventually sweeping away the county structure of the old maps for a simpler grid that largely sliced the country into neat rectangles, reflecting sheet sizes. The practice of using cartography to unite and define disparate territories learnt overseas and elsewhere in the British Empire, was brought back home, at a variety of scales but with a standardised approach and a standard set of symbols. There was little reason for a map of England anymore—it was a concept assumed outmoded. It has, however, proved to be remarkably resilient.

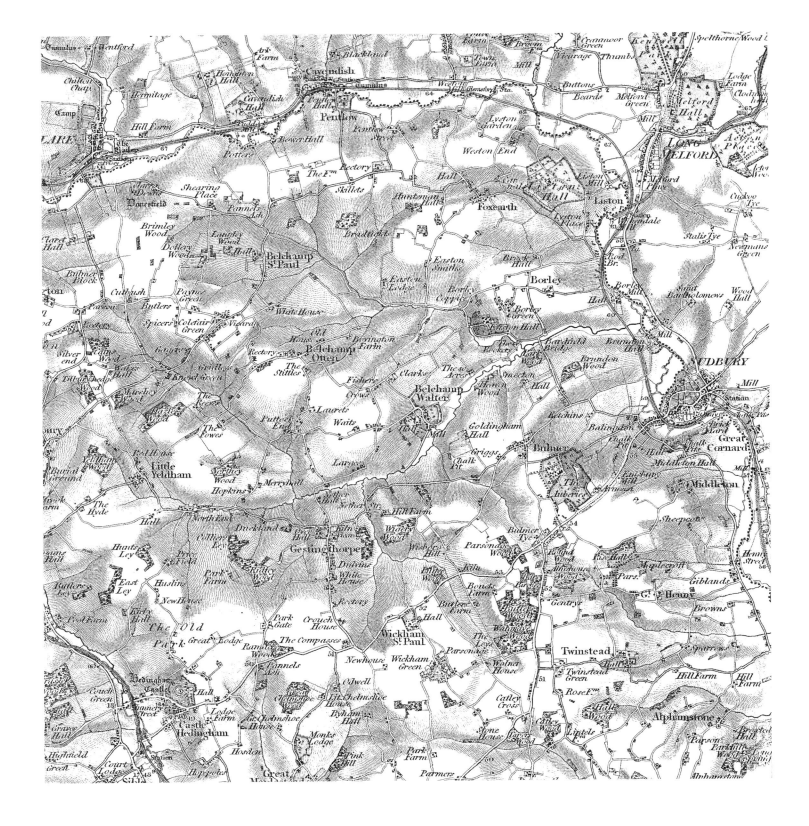

**Ordnance Survey Map of Essex,
First Series, 1805**

A detail of the 1805 one-inch to one mile Ordnance Survey map of Essex, showing four parishes including; Clare, Alphamstone, Little Yeldham and Sible Hedingham. The style has still to settle down into a standardised and recognisable OS format but already many of the more expressionist techniques experimented with in the late eighteenth century have calmed down into a clear communicative style.

Overleaf **Ordnance Survey One-Inch Map of Bournemouth (detail), 1947**

In 1947, Britain's roads were correctly presented for the first time in the fully revised Ordnance maps. The one-inch to the mile map is the most famous of the series of maps to be produced by the Ordnance Survey.

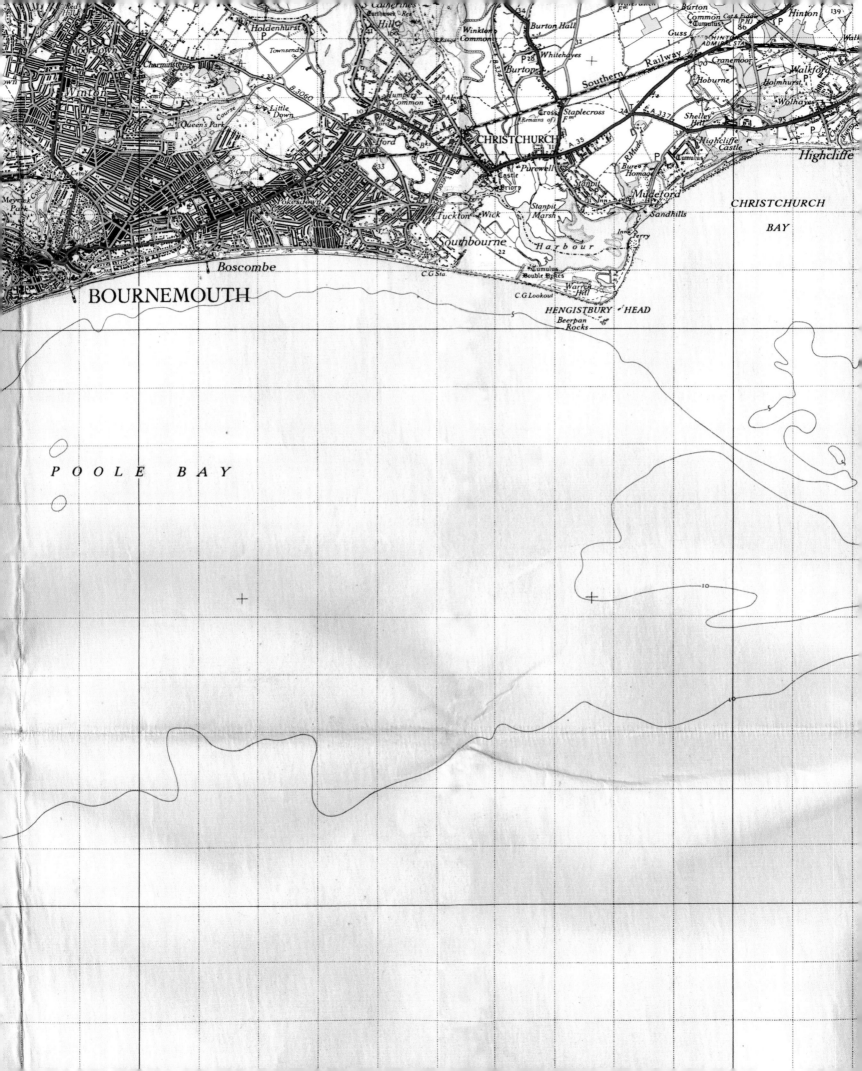

National Grid   179

ORDNANCE SURVEY

DIEU · ET · MON · DROIT

New Popular Edition

ONE-INCH MAP
of ENGLAND & WALES

BOURNEMOUTH

Sheet 179

First
Published
1940

Full Revision 1930
Roads 1947 with
later corrections

○ Ashmore
Fordingbridge ○
Ringwood ○
Wimborne Minster ○
Christchurch
POOLE
BOURNEMOUTH
Wareham ○
Swanage

Price Seven Shillings Net

Front cover of Ordnance Survey
One-Inch Map of Bournemouth, 1947
The recognisable red-covers of the post-war
Ordnance Survey maps, such as this one, were
designed by Ellis Martin. The new popular edition was
first published between 1945 and 1947 in 114 sheets.
It was later replaced by the Seventh Series map, which
was produced from 1952 to 1962 in 190 sheets.

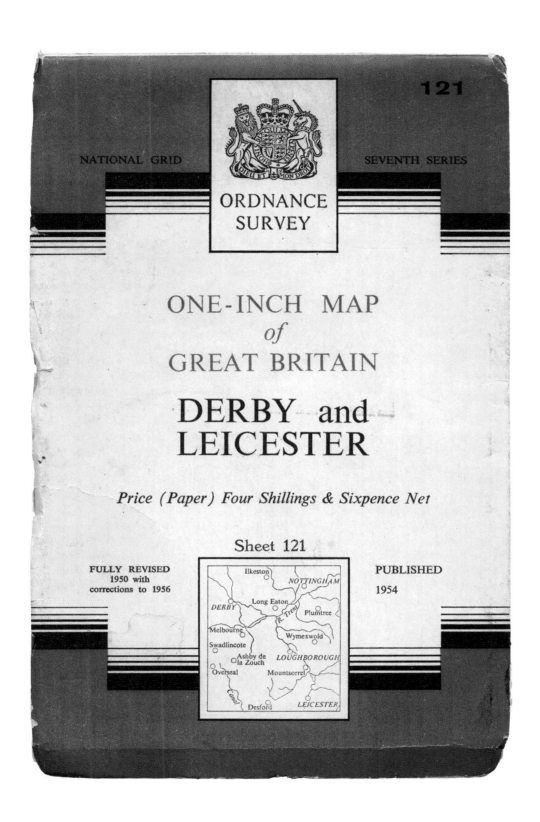

Front cover of Ordnance Survey One-Inch Map of Derby and Leicester, 1954

The covers for the Seventh Series Ordnance Survey one-inch maps, including this of Derby and Leicester, were designed by Ellis Martin in a bid to improve upon the original popular edition of the immediate post-war period. Martin's new cover design were devised to be more attractive to the leisure industry.

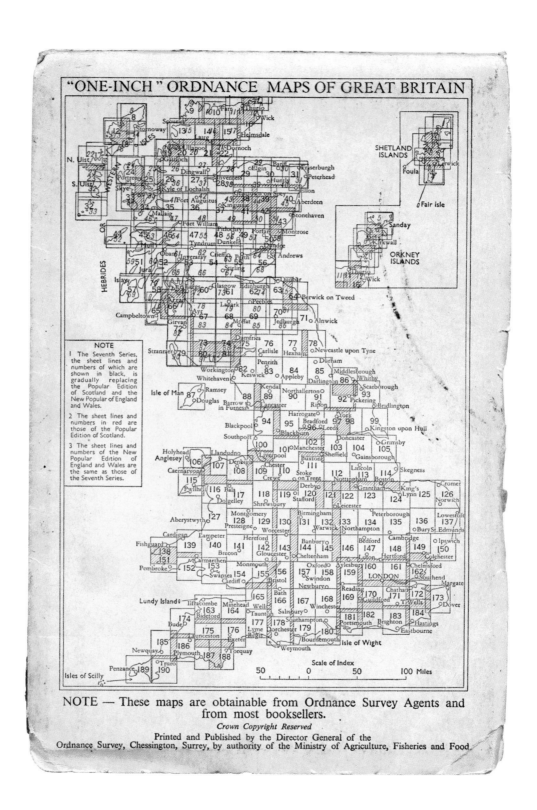

"ONE-INCH" ORDNANCE MAPS OF GREAT BRITAIN

NOTE — These maps are obtainable from Ordnance Survey Agents and from most booksellers.

Crown Copyright Reserved

Printed and Published by the Director General of the
Ordnance Survey, Chessington, Surrey, by authority of the Ministry of Agriculture, Fisheries and Food

Above and Opposite (detail of interior)

Back cover of Ordnance Survey One-Inch Map of Derby and Leicester, 1954

Included on the cover of this map is a description of the National Grid and Referencing System which, when launched, would remain dominant for almost 40 years before being replaced by the 1:150000 scale series.

USEFUL AND INFORMATIVE

DEFENCE OF THE REALM

The association between maps—particularly maps of England—and military planning and intelligence is long-standing. The very title 'Ordnance Survey' reminds one of that association. The importance of maps to the military planner has also guaranteed a funding stream for a particular type of mapmaking that was more elusive for other approaches made with alternative purposes in mind. For this reason, maps that both plan and record attack and defence were often made significantly in advance of others (even if they were frequently kept secret or were unavailable for public use).

Maps made for military purposes are usually practical and single-minded but, as they are almost inevitably produced in periods of heightened threat and insecurity, they can reveal an attitude to national identity sharpened by circumstance. This is at its most pointed on maps produced for propaganda purposes, for example, when depictions of England become synonymous with the English 'values' being fought for and defended.

For centuries, the threat of attack came from only two sources: Scotland and France, with occassional challenges from The Netherlands and Spain. The simultaneous threat from both of these sources during the Tudor period coincided with the rebirth of English mapmaking in the sixteenth century. One of the first examples of this was the series of maps depicting the vulnerable coastlines in the south of England commissioned by Henry VIII, following his divorce from Catherine of Aragon, the split with the Catholic Church and the resultant threat of war with the combined forces of Francis I of France and Charles V, Holy Roman Emperor and King of Spain.

Most of these plans come in a long strip, with the land viewed from the sea. They must have been assembled from the observations of on-ship surveyors and sailors, as well as locals responding to a call in 1539 from Henry's chief minister, Thomas Cromwell, for information. At least one map produced during this period—one of the coastline and rivers of Essex and Suffolk by Richard Cavendish—maps the land more fully while another—of Dorset around the Portland Bill—depicts the sea and the coast from inland. All these maps were intended for use by the King and his advisers at Whitehall, rather than on ship, and were part of Henry's massive and costly effort to fortify the defences along the coastline.

The long views; including those of the coast of Cornwall, Somerset and Dorset are, by their nature, images of England from the 'outside'. By focusing on the land's edge they reinforce notions of containment and separation as well as the country's reliance on the sea and the necessity of naval power for protection. Very little is charted inland, while villages and towns on the shoreline are represented alongside forts and defence installations. By contrast the sea is full of ships, signifying defence and the, increasingly vigorous, trade industry, both local and international.

Dorset Coast, 1539, Anon

Courtesy the British Library

The production of this, largely pictorial, map was
a direct result of the threat of invasion during the
1500s. In 1539, Thomas Cromwell issued a mandate
that required all coastal villages and land to be
surveyed by their owners. As a result he received
a multitude of drawings, sketches and texts that
were later compiled into large roll maps (each
approximately three metres long). It is likely that
this rendering of the Dorset coast was a portion of
one such roll map. As such, this map has a primarily
functional purpose and shows the fortifications,
towers and beacons along this stretch of the south
coast prior to the arrival of the Spanish Armada.

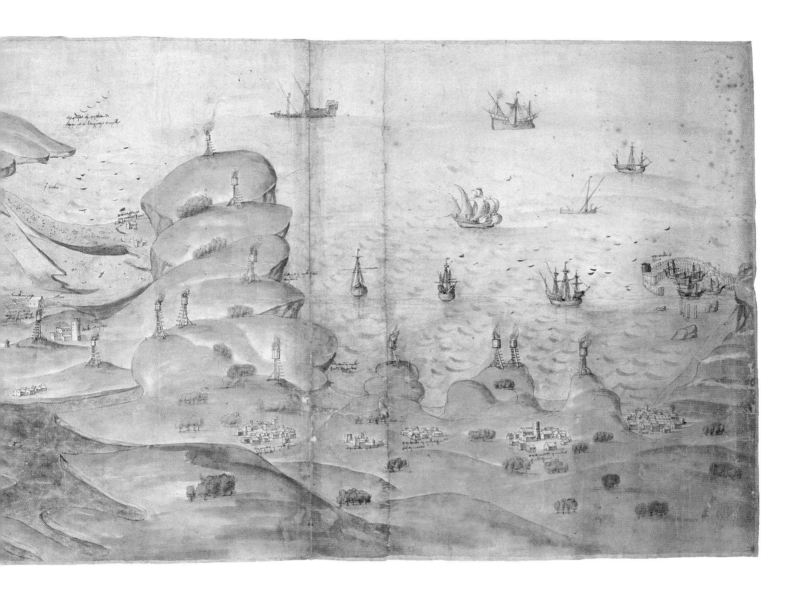

Although the war with France broke out in 1543, the predicted invasion did not take place until two years later, as the French navy sailed up the Solent and temporarily landed on the Isle of Wight. Greater threat to the country's security would materialise in 1588, as the Armada approached from Spain. By that time, the entire country had been mapped by John Rudd, Christopher Saxton and a handful of others—under the patronage and watchful eye of William Cecil. The coast and its rivers, harbours and berths had been fully surveyed and defences put in place. The great atlas of maps that Cecil assembled is now one of the great treasures of The British Library, complete with his copious annotations. His handwritten notes and marginalia mark essential communication routes on Rudd's 1569 manuscript map of County Durham, and fill the edges of his proof copy of Saxton's map of Northamptonshire with information on various Lordships.

The defeat of the Armada by the English Navy, despite the failure of the revenge attack the following year by the 'English Armada', left them in command of the seas surrounding Britain. Similarly, the accession in 1603 of James the VI of Scotland to the throne of England brought an—albeit temporary—end to conflict with Scotland across the border. With such relative stability the need to invest further in mapmaking lost its imperative, and English cartography went into hibernation. When the English Civil War erupted in 1642, both sides were forced to rely on Saxton's 60-year old Wall map of 1583, in a

version engraved by the Bohemian Wenceslas Hollar (still best-known for his post-Restoration maps and views of London). Known as 'The Quartermaster's Map', it was published in 1644 by Thomas Jenner and advertised as "useful for all Commanders for quarteringe of souldiers, and all sorts of persons, that would be informed, where the armies be". Samuel Pepys, an avid map collector, thought highly enough of Hollar's map to give a set as a gift. On 9 June 1667, he records in his diary, just after accusing Lord Chesterfield of going off "to debauch country women.... My Lord Barkeley wanting some maps, and Sir W Coventry recommending the six maps of England that are bound up for the pocket, I did offer to present my Lord with them, which he accepted."[1]

It would take a new kind of civil conflict to awaken English mapmaking from its state of hibernation: the Jacobite rebellion of 1745. Following the Battle of Culloden, and the subsequent brutal suppression of the people of the Scottish Highlands, a complete survey and map of Scotland was ordered by Prince William, the Duke of Cumberland (or 'Butcher Cumberland' as he was otherwise known). William Roy of the Royal Engineers, 1726–1790, was one of the military surveyors employed on this project. Roy would later be able to draw on this experience when starting work on the 'Principal Triangulation' of England, marking a new phase of mapping, both military and civilian. It was Roy who had prepared the ground sufficiently for accurate maps of the southeastern counties to be ready for publication when war once

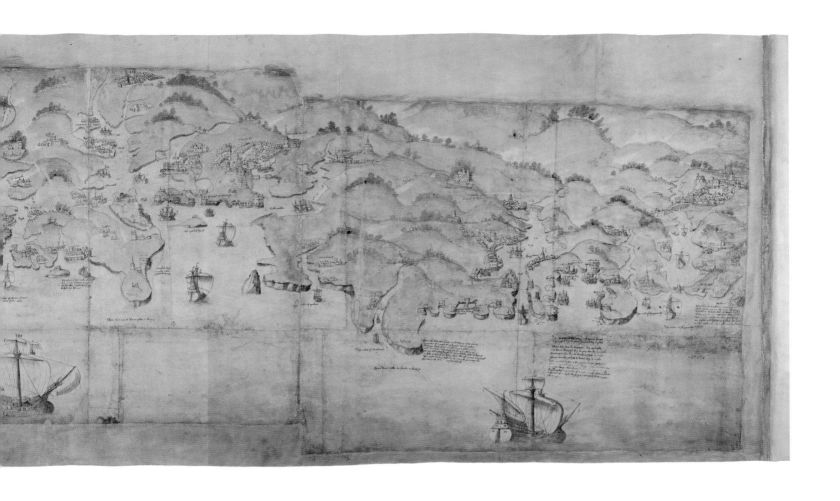

**Southwest Coast of England
from Exeter to Land's End, 1539, Anon**

Courtesy the British Library

This map depicts the southwest coast of England, from Exeter through to Land's End. Despite its scenographic quality, its principal function was to depict potential landing points along the coast. In sections where cliffs would make landing difficult or dangerous, the area is foreshortened. Conversely, areas that made landing easy (such as sandy beaches) are more prominent and visually exaggerated. Although some effort clearly went into the rendering of the coast in order to aid its utilisation by sailors, it was probably never used due to its size —at over three metres long, it would have made for a cumbersome choice at sea.

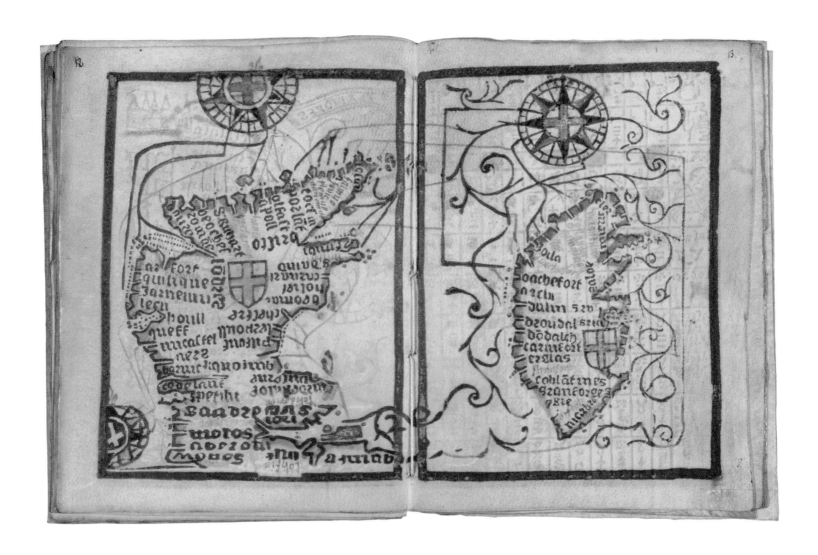

Portolan Chart, 1546, Anon

Courtesy Bridgeman Art Library

This 1546 Portolan Chart originates from France and includes the north coast of Spain, southern Brittany and the north coast of Brittany at Flanders. A rather decorative example of the genre, it depicts two rose compasses, each with a different coat of arms in the centre, as well as the major ports and settlements along the coasts.

again broke out with France, and he is invariably cited as the founder of the Ordnance Survey.[2]

When the Napoleonic Wars commenced in 1803, the Board of Ordnance was already well-established, having published its first map in the one-inch to a mile series of county maps by William Faden. Under the guidance of William Mudge, it had explicitly started mapping those areas of the country most under threat of invasion, working rapidly along the southern coast (Essex, 1805, Devon, 1809, Hampshire, 1810, and so on) so that, by the end of the Napoleonic Wars in 1815, the whole of the Southern part of England had been mapped. Each of these maps was issued at the standard scale of one inch to a mile, although the survey work was initially carried out to a far greater scale (six inches to a mile for areas of military significance with scales of two and three inches to the mile also being deployed).

The title 'Ordnance Survey' first appeared in 1810, but took many years to take popular hold. In the early years, the surveyors were all serving officers, utilising skills based on a long tradition of battlefield surveying. As now, the history of warfare and past battles was a consuming interest for soldiers and laymen alike. This was a passion lampooned by Laurence Sterne in *The Life and Opinions of Tristram Shandy, Gentleman, 1759–1767*, through the character, Uncle Toby, who continually re-enacts the Siege of Namur on the bowling green of Shandy Hall. Similarly, Roy had personally surveyed the Roman Antonine Wall, featured in his book *The Military Antiquities of the Romans in North Britain*, while the Ordnance Survey was meticulous in recording historic monuments and the site of battles as well as publishing dedicated battlesite and historical maps.[3]

British success in the Napoleonic Wars, and the subsequent dominance of the Royal Navy, maintained a more than adequate 'flight distance' from any threat of conflict for almost two centuries. Such was the confidence in the Navy that the First Sea Lord Fisher stated in 1909: "We have got a fleet so invincible that... the worst that could possibly happen to us would be some trouble in India or on the continent which, however disagreeable, could never touch our hearths and homes."[4]

However, this was all about to change with the invention of aerial bombardment. Within six years of Fisher's speech, England would be under sustained attack again. The towns of Great Yarmouth, Sheringham and King's Lynn were all bombed by German Zeppelins during January of 1915, killing four and injuring a further 16 people. Such attacks were mapped in detail by military intelligence, including explosives and damage costs—unemotional records of destruction wrought.

This pattern was carried through to the Second World War with both English and German maps of the country depicting the plans for, and results of, bombing raids. Such maps would become vivid reminders of the effects of war, whose garish, bright colours would suitably express the aggression explicit in bombing from the air.

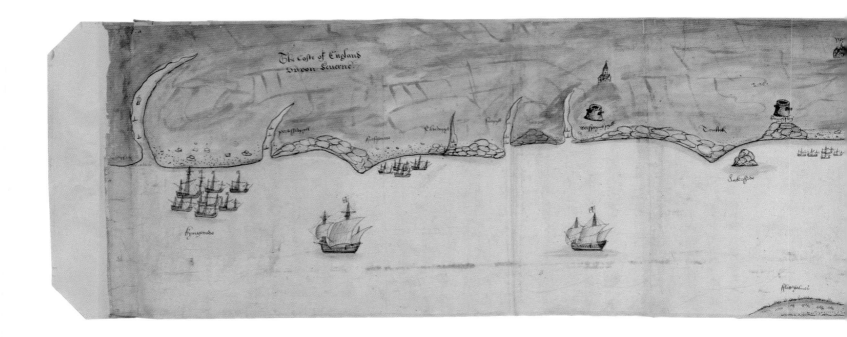

The Second World War also introduced a far more knowing and committed use of propaganda than had been seen before. War had always utilised vituperative images to sway public opinion, including those such as Charles William's 1803 Bone of Contention map (in which Napoleon is depicted as a monkey and John Bull, a bulldog), but the contrast between the images evoked by the two sides is marked. While German propaganda presented England as a threatened land, the home government depicted the country as a bucolic ideal (even while ploughing up of land was encouraged to stave off starvation).

The Soviets seem to have eschewed such crude imagery but were, nonetheless, committed mappers of England; producing a variety of maps during the Cold War that only became readily available after the fall of the Iron Curtain. Such maps were developed at a range of scales and drew on whatever information could be gathered from public and clandestine sources. They are startlingly accurate and practical even if the militaristic style and Cyrillic renaming of the country is slightly chilling.

A Coloured Chart of The Coast of England upon Severne being the Whole North Coast of Somersetshire; with the Forts erected thereon, circa 1540, Anon
Courtesy the British Library

This pictorial representation of the north coast of Somerset includes the river Avon near Bristol and, as with many pictorial maps of the coast from this time, was made in response to the threat of invasion. This is apparent both from the map's exaggerated rock cliffs and outcrops (an indicator of England's 'impenetrability') and the fact that political upheaval was particularly rampant during the period in which it was created. The vision of England as an impenetrable realm is further enhanced through later additions to this work, which included fortifications, towers and gun platforms.

ESSEX

KENT

SVSSEX

This scale in length doth contayne 20 myles

A Coloured Chart of the Course of the Rivers Thames and Medway, and of the Coasts of Kent and Sussex to Shoreham, with an Account of the Tides, 1596, William Borough

Courtesy the British Library

This map communicates the pressing concern of how to effectively defend the River Thames and London, which were continually under threat during the Anglo-Spanish War. The map was carefully rendered and the coastline appears to have been tinted by hand. Though it is not signed, the script is most likely that of William Borough, a British naval officer.

Overleaf **Northumbriae Comitatus, circa 1570, Christopher Saxton**

Courtesy the British Library

A proof copy of Christopher Saxton's county map of Northumberland owned and annotated by William Cecil and produced for Saxton's Atlas of England and Wales (published in 1579). Saxton's work, which accurately surveyed and mapped the entire country, was commissioned and financed by the Master of Requests at Elizabeth's court, Thomas Seckford, although the major motivating force behind the project was Cecil, the Secretary of State (who was determined that the whole country should be mapped and himself was an avid collector of maps). Cecil believed in the power of information, particularly concerning a nation for which he had direct responsibility. The production of accurate and high quality maps, particularly those of borders and of vulnerable coastlines, was a matter of great interest to him.

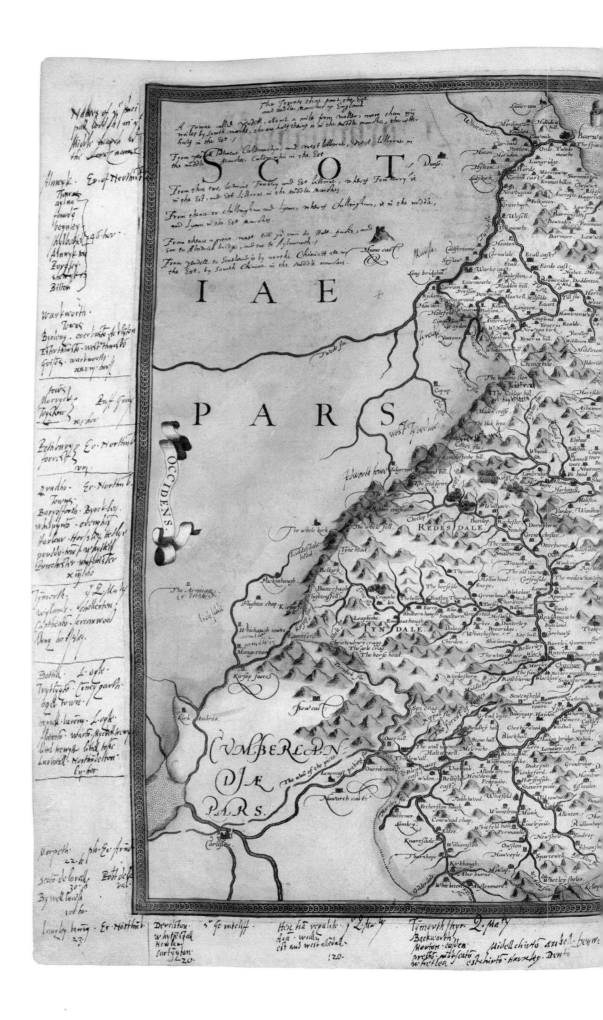

Map Showing the Route of the Armada Fleet, 1588, Robert Adams

Courtesy Bridgeman Art Library

This sixteenth century engraving by Augustine Ryther, appropriated from an image by the artist, Robert Adams, depicts the conflict between the Spanish and English naval fleets. It shows the route of the Armada during the Spanish quest for invasion in a dramatic and dynamic—if not wholly accurate—manner. Ryther also engraved many of Chrsitopher Saxton's county maps but does not appear to have transferred his knowledge of English topography to this image.

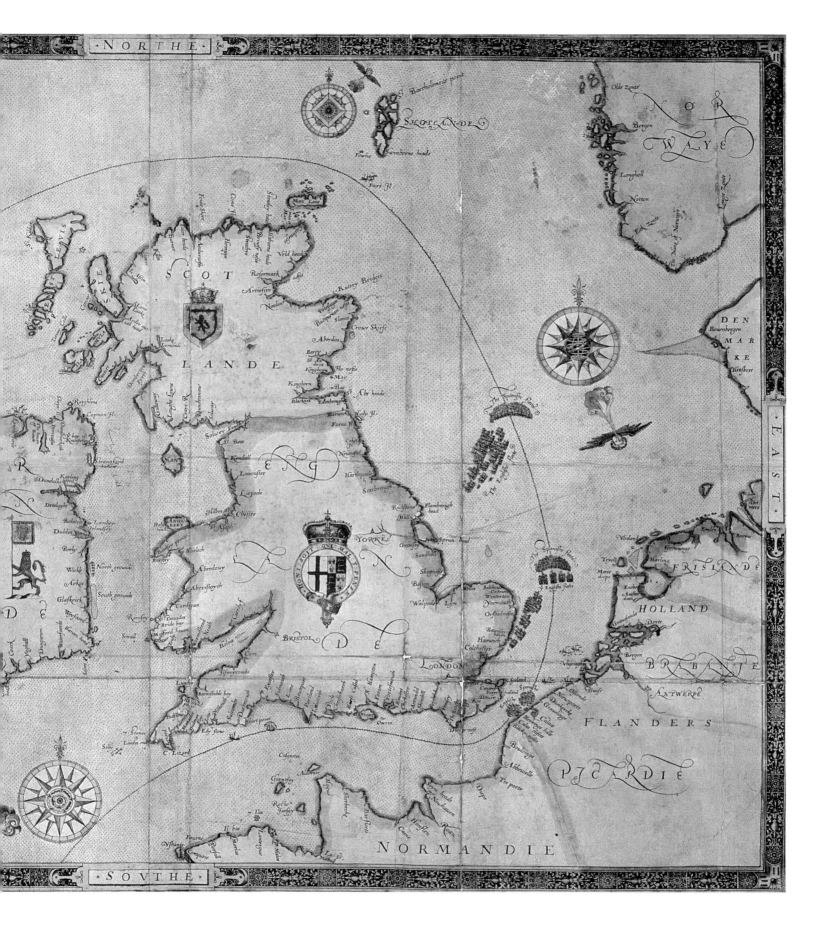

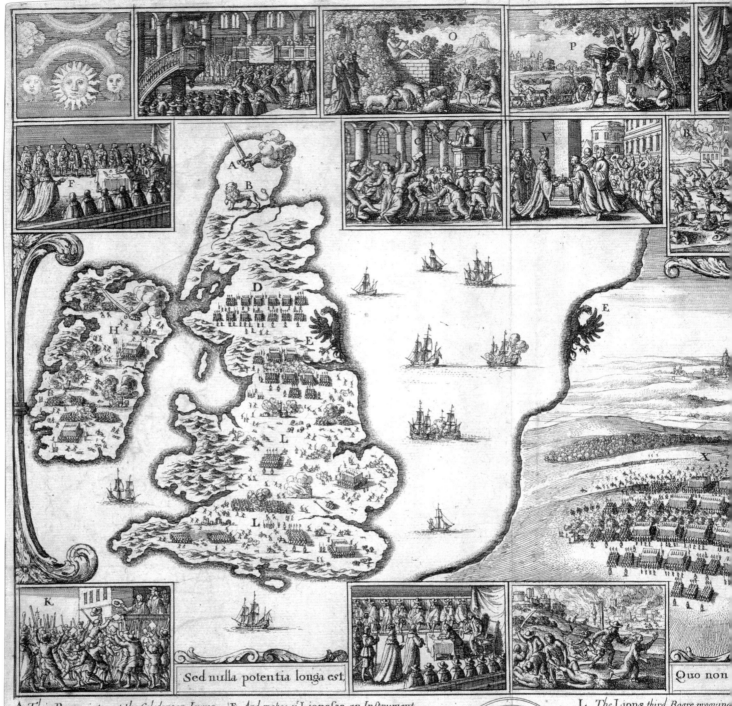

Sed nulla potentia longa est

Quo non

at Cæfar to y Romans *Crown doth bring*
ftately Nephew *and creates him* king.
que giues y Crown to Frederick *& excites*
Sword to affert the Germans, & their Rights
n from high Windowes, unnawares were thro
Emperors *Councell ere the Charge was known,*
Blow neer Prague *was ftruck. The people ride*
Ichu out, Warre is fweet before t is tryd
t Decollations then? What Blood? What fair
cted Tragick Scenes enfud that Warre.

**Map of Civil War England and a view
of Prague, 1632, Wenceslaus Hollar**
Courtesy Bridgeman Art Library
In Wenceslaus Hollar's map, both England and
Bohemia play prominent roles. Largely pictorial, this
piece probably had personal resonance for Hollar
who, although living and working in England, was
Czech by birth. The Civil War this map alludes to
greatly impacted upon his trade and he lost all his
finances as a result (although he would later receive
the occasional commission from the Duke of York).

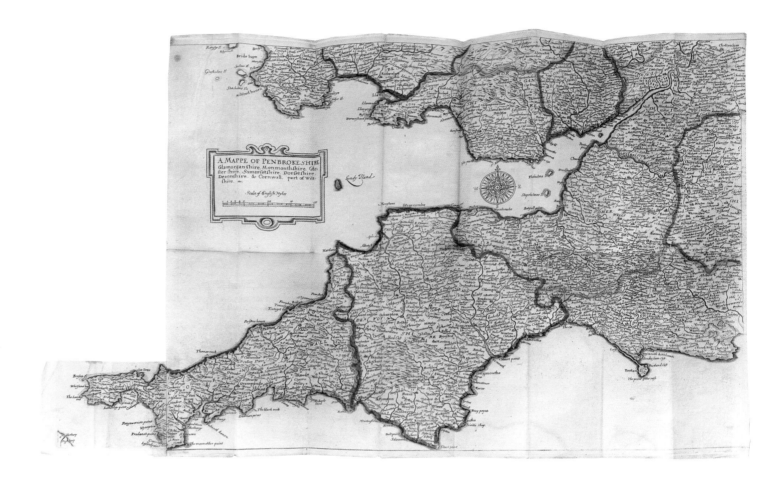

**Quartermaster's Map, 1676,
Wenceslaus Hollar**

Courtesy Bernard Shapero Rare Books
Published by Thomas Jenner in 1644, this compilation
of six maps (with a key map and a separate sheet
containing three smaller maps) were engraved
by the prolific Bohemian artist and cartographer,

Wencelaus Hollar. The map was a reduced and re-
engraved version of Christopher Saxton's renowned
large wall map of 1583, *Brittania: Insurarem in Oceano
Maxima*. Jenner and Hollar's version was designed in
the same format, but reduced in size, so that sheets
could either be mounted together on a wall or
bound and used as an atlas. Advertised as "Vseful for

all Comanders for quarteringe of souldiers, and all
sorts of persons, that would be informed, where the
Armies be", it subsequently became better known as
the Quartermaster's Map, and was used by both the
Royal and Parliamentary Forces during the Civil War.

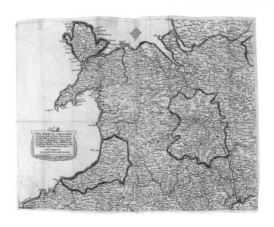

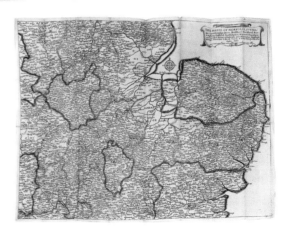

The Bone of Contention or the English Bull Dog and the Corsican Monkey, 1803

Courtesy Mary Evans Picture Library

There is a long history of using maps in caricature and cartoon. Icons often used as shorthand for nations themselves—particularly during periods of war and conflict—they communicate the longstanding popular understanding of the 'national map' as an identifier for the nation-state.

The Bone of Contention shows a Corsican monkey and an English bulldog fighting over Malta. In this caricature William Charles depicts Napoleon as the monkey and John Bull as a bulldog both standing on a large invasion plan for England, on which markings show a list of boats, covering the Channel and the North Sea.

**"Come lad, slip across and help",
recruitment poster from the First World War,
published by The Parliamentary Recruiting
Committee, London, 1915**

Courtesy Topham Picture Source/
Bridgeman Art Library

This cartoon, produced by the Parliamentary
Recruiting Committee in 1915, was one of many that
were deployed in the bid to recruit soldiers during
the First World War.

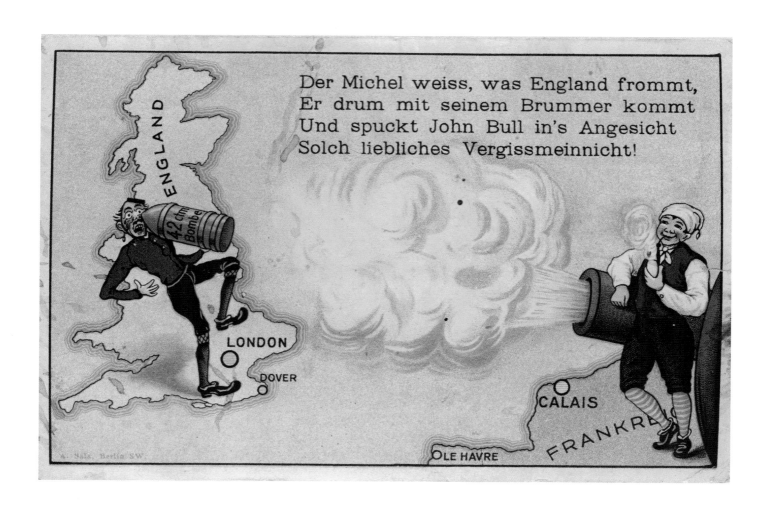

German Propaganda Postcard, circa 1916

Courtesy NMeM *Daily Herald* Archive/
Science & Society Picture Library

At the time of the Third Reich Germany's strength became manifest in its *kultur* (or culture), which depended on loyalty to German values and systems of governance. Derived from this was a sense of superior morality and general national pride. In the propaganda postcards that proliferated at the time, enemies of the country were frequently ridiculed. This is evidenced in the portrayal of 'Michel', a French fictional figure who wears a sleeping hat rather than a uniform. 'John Bull' is another such example, a national personification of Britain, he was often depicted as ineptly leading his country to battle.

Britannia 'high and dry', blockaded by the U-Boats in this First World War Cartoon, 1917

Courtesy Mary Evans Picture Library

In this First World War, cartoon Britain floats precariously above water. Churchill stands on guard at every prospect, a clear signal to the ensuing threat of the Axis Powers. A satrical caricature, Churchill stands casually reading the paper, scratching his head or looking off into the distance. The title clearly makes reference to the potential invasion of Britain by German U-Boats (or underwater boats), motivating Churchill's removal of the nation from the ocean in this cartoon. The large fish positioned in the south of the country implies that while the nation might be safe from German invasion, the presence of the Japanese forces could clearly be felt.

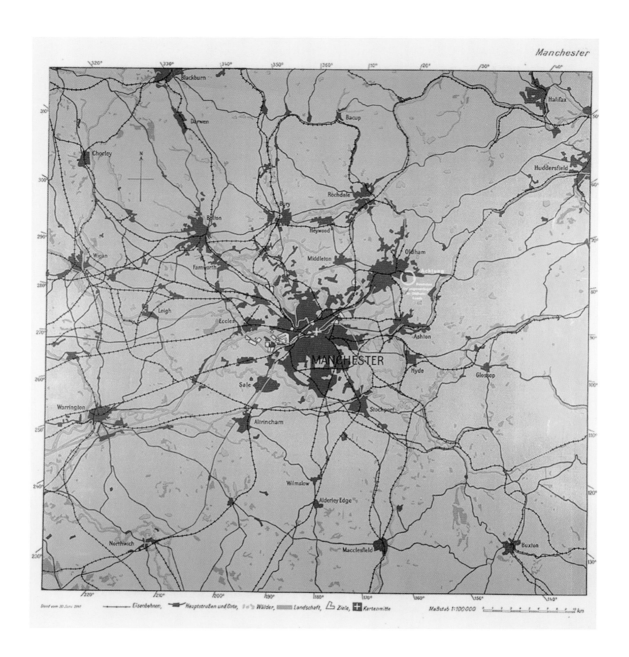

Above **German Bombing Map**
of Manchester, 1941

Luftwaffe bomber pilots would often possess a series
of such maps as this view of wartime Manchester, in
order to aid their nighttime missions over Britain.
The topological features of these maps had to be
obvious—towns, railways, major roads and wooded
areas were always present. However, there were
some areas to be avoided; one of these, just outside

Oldham, has a yellow circle around it. Marked
with the words "*achtung: Deutshes gefangenenlager*
in Oldham—Leeds", it alerts pilots to a German
prisoner of war camp. Its bright colours aided pilots
in identifying details at night, the red and yellow
sections being the target areas for the mission set.

Opposite **Targets in the British Isles**
from *Signal Magazine*, no 3, 1941
Courtesy Bridgeman Art Library
A 1941 lithograph of the Nazis' 'targets' in Britiain.
Originally published in *Signal Magazine*, it includes
military ports, building sites, garrisons, oil and corn
stores, iron and coal mines and aluminium factories.

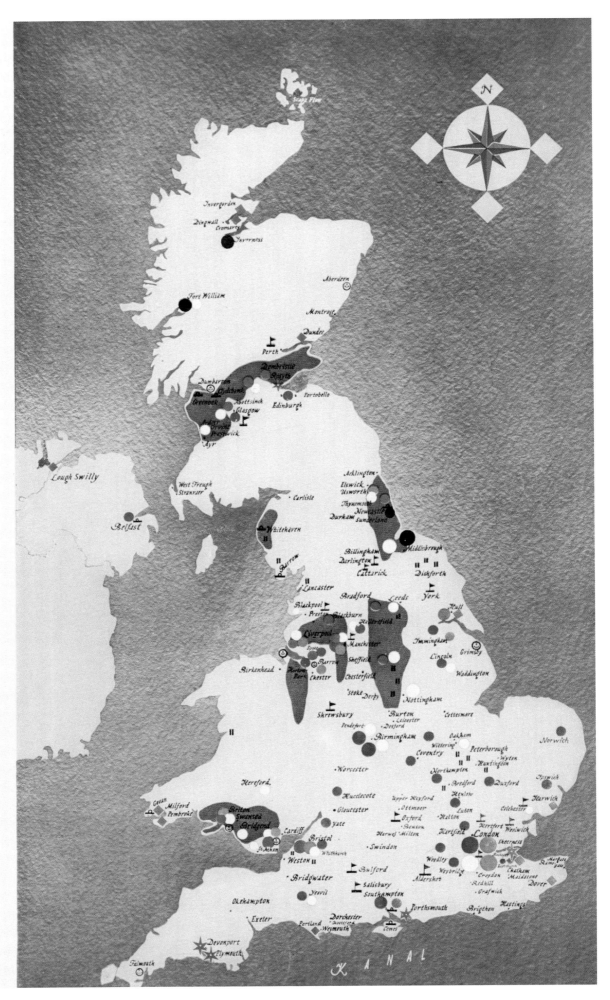

Des buts sur l'île d'Angleterre

Cette carte d'Angleterre montre les buts qui ont été exposés, jours et nuits, aux puits des bombes allemandes, et qui le seront encore : des ports militaires, des chantiers, des garnisons, des stocks d'huile et des silos de blé, des usines d'armement, des mines de charbon et de fer, des usines d'acier et d'aluminium. Pendant que la marine de guerre allemande coupe toute communication et que l'encerclement de l'île se fait de plus en plus efficace, l'arme aérienne atteint les autres centres vitaux de l'adversaire à l'intérieur de l'île, qui sont d'une première importance pour la guerre et qui sont répandus sur le pays entier

Port militaire,
1ère catégorie

Port militaire,
2ème catégorie

Des chantiers, où des navires de guerre se trouvent en construction

Port de première
catégorie

Des divisions
(garnisons)

Des distributeurs d'essence (huile minérale)

Des usines d'avions

Des silos de blé et des
moulins

L'industrie de matières
premières, des mineries
de fer, des usines d'acier
et d'aluminium

Des usines d'armement
y compris des produits chimiques

Des districts houillers

Des gisements de fer

Map depicting the "Ploughing Up Campaign"
of England and Wales, circa 1940s
Courtesy Mary Evans Picture Library
This map shows the extent of the Ploughing-Up
Campaign in England during the 1940s. A response

to extreme food shortages during the Second
World War, the underlying concept of the campaign
was to maximise farm produce. The Ministry of
Agriculture initiated the campaign as part of the
'war effort', and it came into action at remarkable

speed and scale shortly thereafter. In this map, there
is a reference to show the area of crops in 1939 and
another depicting the increase of crops between
1939 and 1942.

The Ploughing-up Campaign in
ENGLAND & WALES

Represents area of crops in 1939

Represents increase of crops between 1939 and 1942

Represents built-up areas

SCOTLAND

NORTH SEA

IRISH SEA

ENGLISH CHANNEL

1	NORTHUMBERLAND	16	NOTTINGHAM	31	WARWICK	47	HERTFORD
2	CUMBERLAND	17	CAERNARVON	32	NORTHAMPTON	48	ESSEX
3	DURHAM	18	DENBIGH	33	HUNTINGDON	49	SOMERSET
4	WESTMORLAND	19	MERIONETH	34	ISLE OF ELY	50	WILTS
5	YORK, NORTH RIDING	20	SHROPSHIRE	35	SUFFOLK, WEST	51	BERKS
6	YORK, WEST RIDING	21	STAFFORD	36	SUFFOLK, EAST	52	MIDDLESEX
7	YORK, EAST RIDING	22	LEICESTER	37	PEMBROKE	53	LONDON
8	LANCASHIRE	23	RUTLAND	38	CARMARTHEN	54	SURREY
9	LINCOLN, LINDSEY	24	SOKE OF PETERBORO	39	BRECKNOCK	55	KENT
10	LINCOLN, KESTEVEN	25	NORFOLK	40	GLAMORGAN	56	CORNWALL
11	LINCOLN, HOLLAND	26	MONTGOMERY	41	MONMOUTH	57	DEVON
12	ANGLESEY	27	CARDIGAN	42	GLOUCESTER	58	DORSET
13	FLINT	28	RADNOR	43	OXFORD	59	HAMPSHIRE
14	CHESHIRE	29	HEREFORD	44	BUCKINGHAM	60	SUSSEX, WEST
15	DERBY	30	WORCESTER	45	BEDFORD	61	SUSSEX, EAST
				46	CAMBRIDGE	62	ISLE OF WIGHT

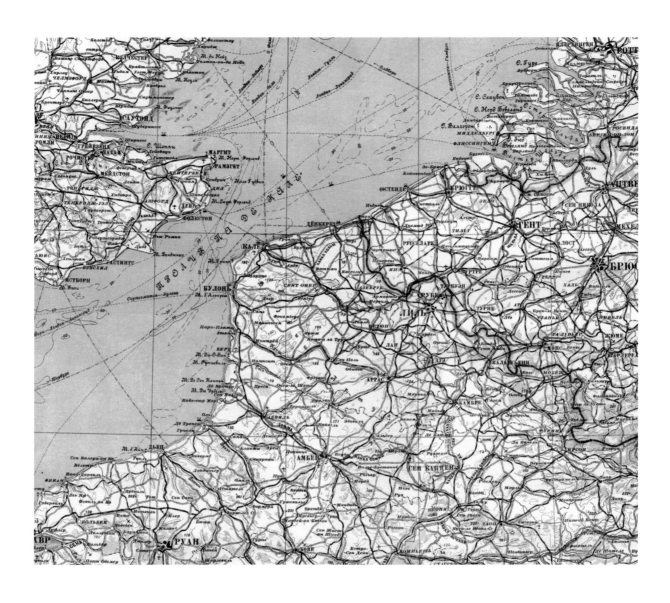

Above **Soviet Map of the English Channel, 1939**
Opposite **Soviet Map of the English Channel, circa 1970**

Russian spy maps of Britain have only been recently released for public access. Produced by the KGB during the Cold War, they were intended to act as preludes to the Russian domination of the world. This map includes topological features not present in the Ordnance Survey maps of the time because of military and political sensitivities. Such maps were compiled using aerial photography, satellite images and local knowledge. The Russians succeeded in mapping 6,178 square miles of the country, and 103 major British cities and towns. The two maps that we see here were compiled, respectively, before and during the Cold War.

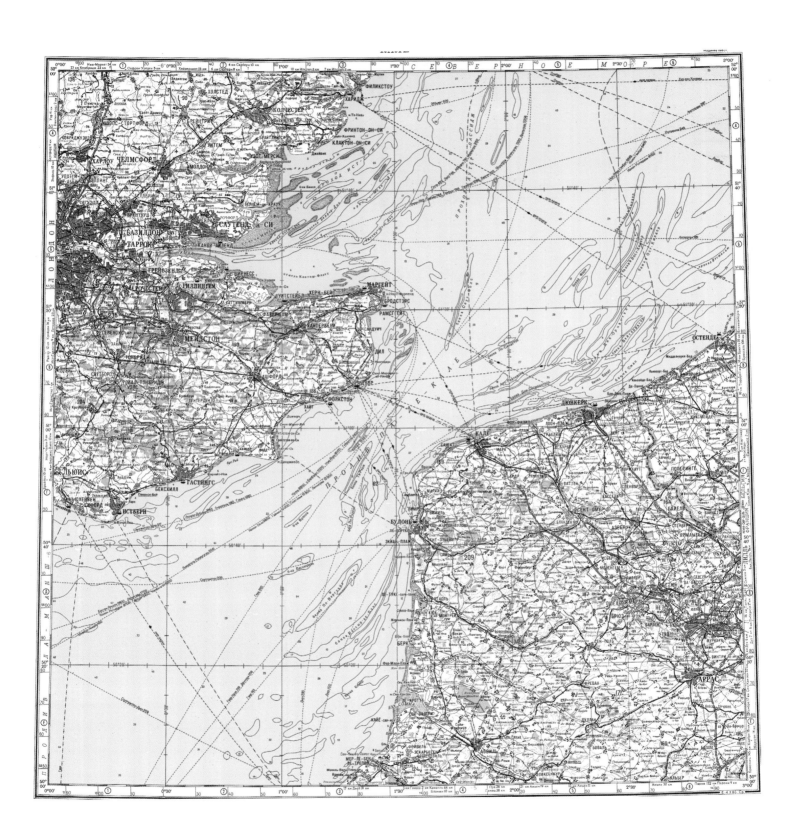

**Soviet Map of Northwest England,
circa 1975**

The Russian comprehensive survey, extending not just to Britain but across the world, was highly detailed and accurate. The exact location and purpose of all structures that might appear to be of a military connection were mapped. The information the KGB compiled was thorough and included the width of roads, the height, width and carrying capacity of bridges, the depth of rivers, locations of train and bus stations and prisons. In addition, there were at least 80 British urban areas mapped and all towns had a comprehensive street gazetteer with a description of the locality and a list of important buildings. Structures of strategic importance are depicted in great detail: military establishments are green, administration buildings purple and industrial buildings black. London, Birmingham, Liverpool, Manchester, Leeds, Sheffield, Newcastle, York, Edinburgh and Glasgow were all mapped in this way.

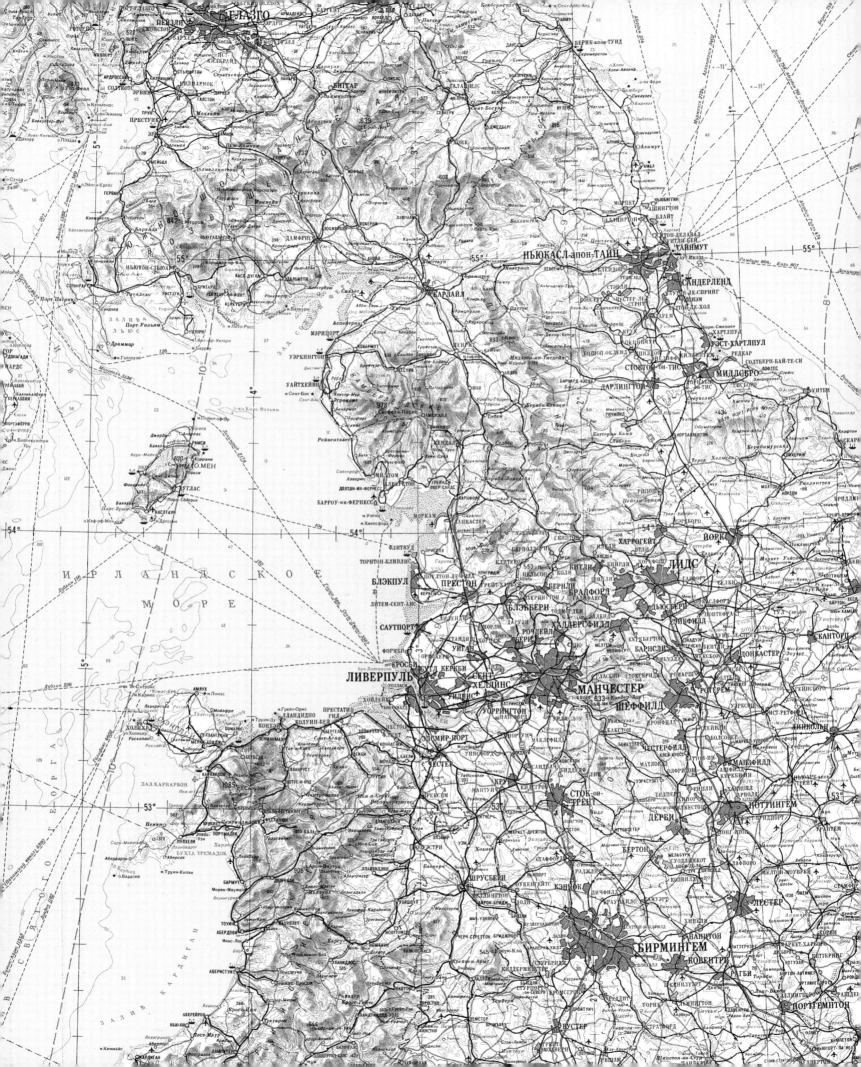

TOPOGRAPHY AND LAND USE

If there is one single purpose of a map, it is to describe a specific land. England is a land particularly heavy with history and cultural value, often best epitomised in literature (such as the fourteenth century description in *Gawain and the Green Knight* of "mist muged on the mor, malt on the mountez" to the Malvern Hills of Elgar, the "Four thousand holes in Blackburn, Lancashire" or Ian McEwan's *On Chesil Beach* of 2007).[5] Maps describing the topography of the country equally command a particular essence of 'Englishness': "If one wanted to show the foreigner England, perhaps the wisest course would be to take him to the final section of the Purbeck hills, and stand him on their summit, a few miles east of Corfe. Then system after system of our island would roll out under his feet."[6]

How can the rolling topography of England, the dramatic landscape of lakes and moors or those "blue remembered hills", be described?[7] Hills and mountains have been prevalent on maps since some of the first made, but it was only when surveying became a more precise science, and as the fever of canal building took hold in the late eighteenth century (following the completion of the Bridgewater Canal in 1761), that the mapping of these romantic and picturesque features became an essential commercial concern.

ALTITUDE PROBLEM

As Catherine Delano-Smith and Roger Kain state in their essential and authoritative study, *English Maps: A History*, "where the eighteenth century maps failed was in the representation of the third dimension of the landscape and in their measurement of altitude". (They continue by citing Thomas Jeffreys' 1771–1772 survey of the Yorkshire peaks—the heights of which he quoted in yards instead of feet—as an example.)[8]

The first of the county mapmakers to tackle such a "third dimension" was William Yates on his 1786 map of Lancashire. Yates used the technique of hachuring to indicate areas of upland relief, borrowing from such maps as John Ogilby's cartographic recordings of roads in the 1670s. Such a technique implied a graphic impression of the shape of the land and was experimented with by many subsequent mapmakers. Some of the most innovative examples include Isaac Taylor's map of Dorset, 1765 and, although rules for hachure work were formalised by the Austrian Johann Georg Lehmann in 1799, the military surveyors of the early Ordnance Survey carried on experimenting long into the nineteenth century.

As such, they reached a degree of artistic apotheosis in the 1820s and 30s with the combined hachuring and tonal shading work of Robert Dawson (who was described by the fledgling Board of Ordnance in 1805 as one of their four "first class surveyors and draftsmen").

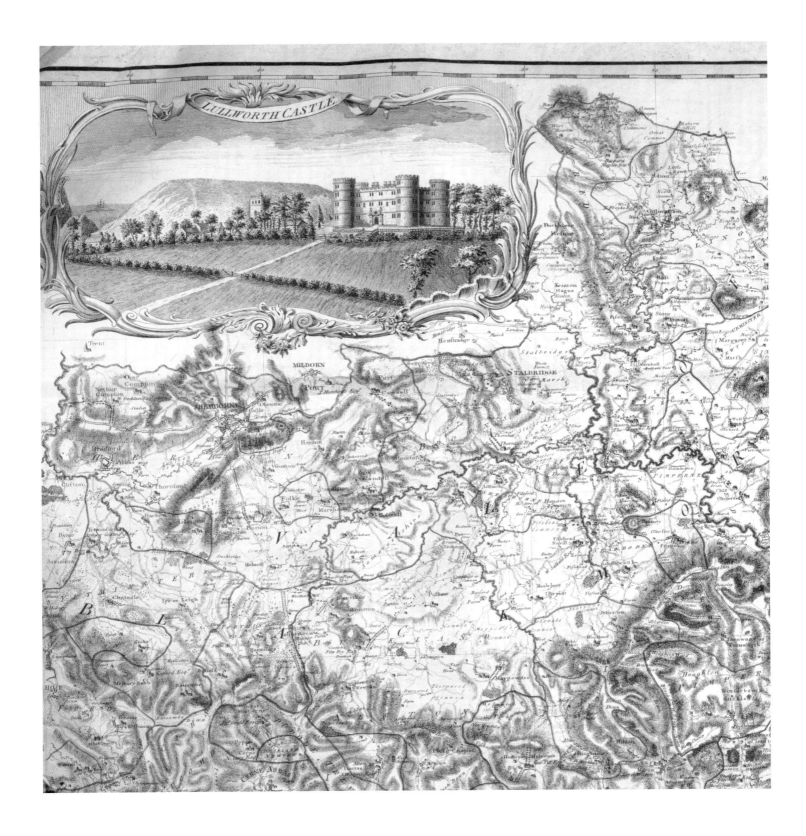

Dorset, 1765, Isaac Taylor

Courtesy Ashley Baynton Williams

Isaac Taylor was a surveyor and cartographer from Ross, Herefordshire. His best-known map is that of Dorset at one-inch to one mile, which he took to engraving himself after being dissatisfied with the work undertaken by other engravers on his earlier map of the county. Taylor's style is richly detailed and he employs a highly wrought freehand hatching technique to express the topography of Dorset which—with its elaborate coastline and iron age hill forts—gave him plenty of scope to indulge in. Taylor submitted this map for the first Society of Arts premium but his technique was judged badly and the award was given to Benjamin Donn for his map of Devon.

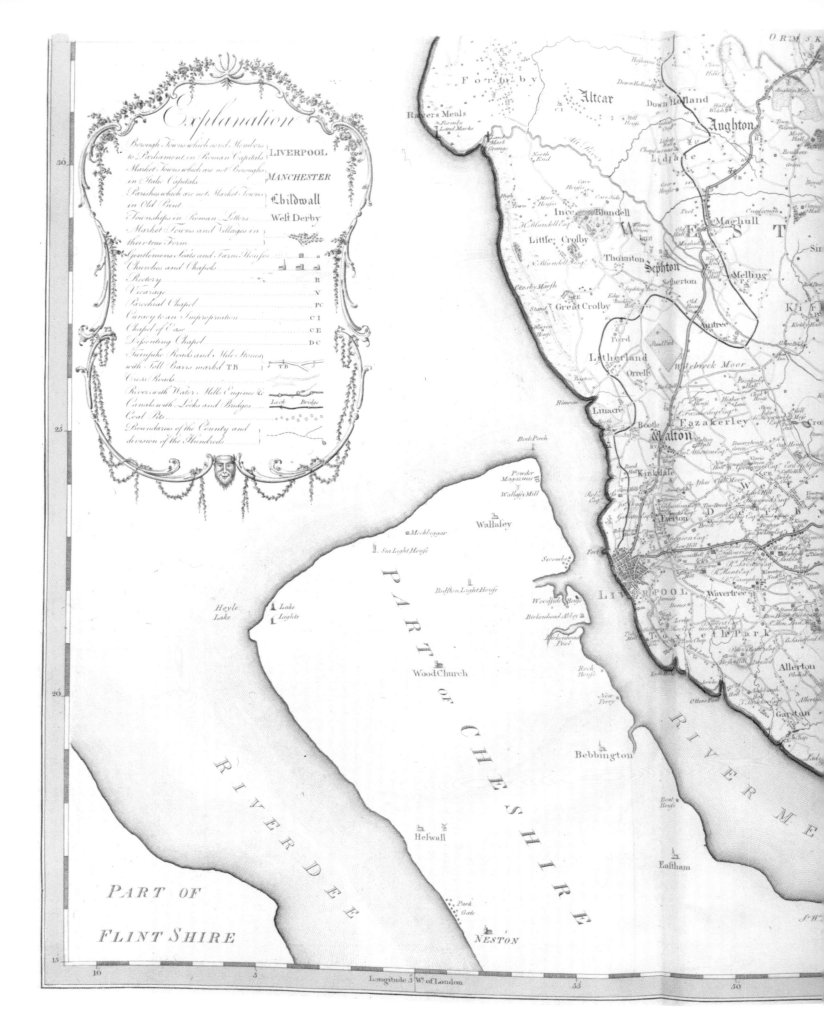

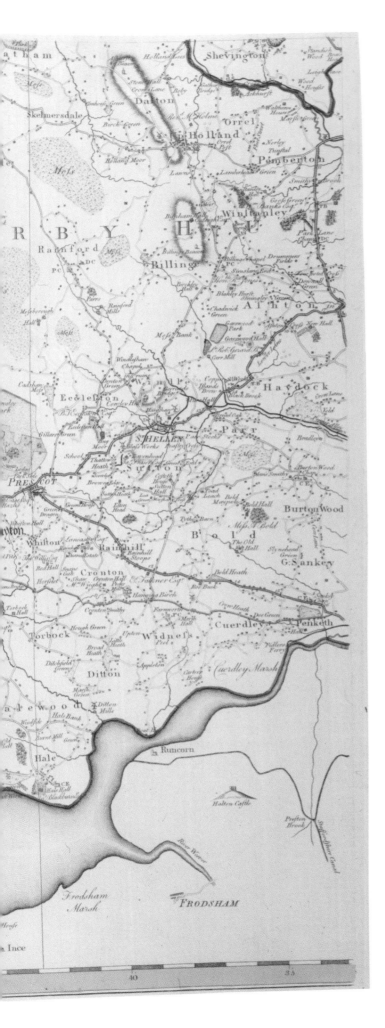

Lancashire, 1786, William Yates

Courtesy Ashley Baynton Williams

Described by JB Harley as the "culmination of pre-Ordnance Survey cartography" William Yates' map of Lancashire at one-inch to the mile won him a coveted Society of Arts award. Drawn and engraved prior to the adoption of contours he uses hatching expressively to indicate the edges of hilly areas rather than attempting to describe the shape or topography of the land. Yates used the known, triangulated, position of St Nicholas's Church in Liverpool to fix his work in the context of the national 'map'.

However aesthetically pleasing and precise the techniques of hachure and tonal relief work were, they were of limited use to those working the land in more practical ways. The eventual solution—contours: lines joining points of equal height—were first used in the province of Cruquius, The Netherlands, in 1729 to measure the depth of the Merwede Canal and were later adapted in 1737 to record the depth of the English Channel. Charles Hutton, professor of mathematics at the Royal Military Academy, Woolwich, would be the first to apply the technique to record the landscape in an attempt to calculate the volume of the Schiehallion mountain in Perthshire. (He apparently "fell upon" the method of "connecting together by a faint line all the points which were of the same relative altitude".)[9] Although contours were utilised in surveying from as early as 1790, it was not until 1843 that they appeared on Ordnance Survey maps as a matter of practice. In the meantime, canal surveyors relied on sections drawn through surveyed areas of land to calculate the best routes for their necessary level works and the optimum location for level changing locks.

Following the opening of the Bridgewater Canal in 1761—and before the emergence of the railway system in 1840—the English went wild for their country's canals. Canal 'mania' really took hold at the height of the Industrial Revolution, and would lead to the construction of over 4,000 miles of man-made waterways, with all the extensive engineering, locks, aqueducts, tunnels and boat lifts that went along with them.

WHAT LIES BENEATH

William Smith, the son of an Oxfordshire blacksmith, was appointed Surveyor and Engineer for the Somerset Coal Canal in 1794, a double-branched canal leading from the mining area of Radstock to the Kennet and Avon canal at Limpley Stoke. Smith, already familiar with the Radstock mines, had begun to develop an understanding of the very new and undeveloped discipline of geology; and the horizontal cuts of the canal into the rocks below the surface revealed to him the sloping structure of the different layers, or strata, of the rocks: "I observed a variation of the strata on the same line of level and found that the Lias rock which, three miles back, was a full 300 feet above this line, was now 30 feet below it, and became the bed of a river and did not appear any more at the surface."[10]

From this realisation—that the rocks below ground were arranged in layers in a fixed pattern and inclined at an angle so that the rock at the earth's surface was constantly changing—Smith knew it was possible to create a map of the underlying geology of the whole of England. Creating such a map became an overriding obsession, driving him deep into debt (and, subsequently, the notorious King's Bench debtors' prison in South London) as well as bitter arguments with the newly established Geological Society of London. Smith would go on to produce a map that charted evidence of fossils in England's geology: evidence that would support Charles Darwin's later theory of evolution. These map can justifiably be seen as providing the research basis for a scientific and theological revolution that shook the world.

Roger Osborne, in his book on the early days of geology, *The Floating Egg*, suggests that this map was the result of an epiphany Smith experienced as he surveyed the landscape from the top of York Minster Tower, while on a canal study tour in 1794. In truth, however, his map was the painstaking work of largely solitary travels across England.[11] It was a monumental task and he was the very last surveyor to carry out anything as ambitious. In their final form, his findings were engraved onto base plates by John Cary but, even then, Smith did much of the colouring himself, using different watercolour washes to represent the varying rock types and strata.

At first, The Geological Society excluded Smith from their ranks and even borrowed some of his survey work for a map of their own. In 1831, however, they repented and awarded him the first of their prestigious Wollaston Medals in recognition of his work and position as "the father of English geology".[12] A learned society was recognising, perhaps rather after the event, that something profound had occurred both to and through their discipline. Mapmaking had moved beyond the depiction of the immediately visible and functional, beyond the antiquarian and aesthetically pleasing, to explore areas of knowledge less immediately apparent. Smith's *A Delineation of the Strata of England and Wales* of 1815 was the first map of a new type of cartography, one that dealt with the representation of different categories of knowledge in map form. It was the start of a journey that would initially encourage the plotting of many different types of data on spatial co-ordinates, but it would also ultimately lead to the liberation from strictly spacial co-ordinates altogether.

In 1835, the Ordnance Geological Survey was set up within the Board of Ordnance (eventually becoming the British Geological Survey in 1984). They would became the publishers of some of the most authoritative geological maps produced of England and still publish—albeit an adapted version—Smith's great map.

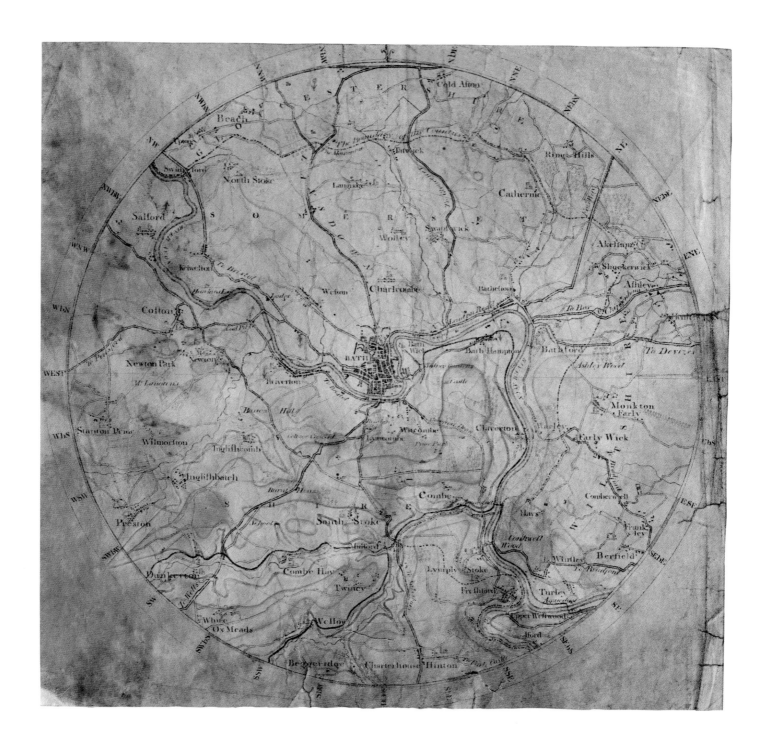

Geological Map of the Neighbourhood of Bath, circa 1799, William Smith

© Oxford University Museum of Natural History/ Bridgeman Art Library

William Smith was an engineer working on the Somerset canal system at the end of the eighteenth century. He was fascinated by the geology of the local land and by the fossils of both fauna and flora found in the rock samples, helping him to identify and begin to order the different geological strata. This map of the area around Bath is his first attempt to map what he observed below ground. It is the first geological map, and a trial for his renowned map of the whole of England and Wales, published 16 years later.

Smith based this map on one of Bath from the guidebook *The Historic and Local New Bath Guide*, printed by A Taylor and W Nayler. He coloured his rock strata over the black and white original— colours that have somewhat faded with age.

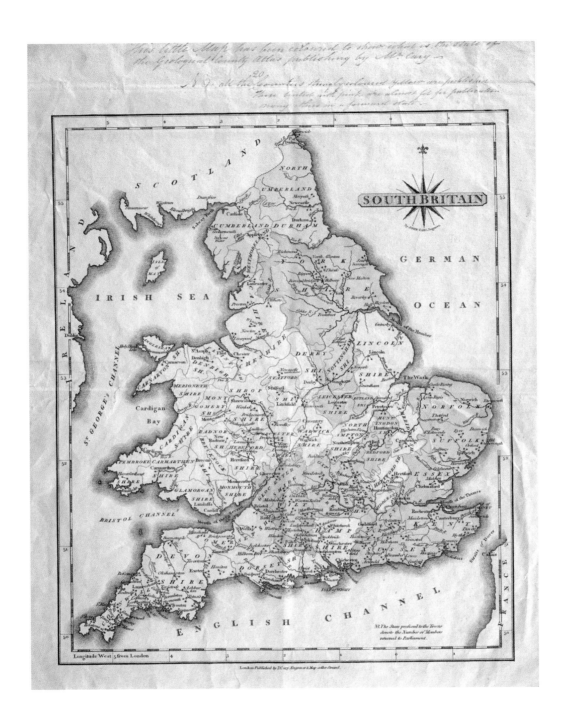

Map of Southern Britain, Coloured to Show Geological Areas of Interest, circa 1790, William Smith

© Oxford University Museum of Natural History/ Bridgeman Art Library

An early study by William Smith, this map demonstrates his speculations on the locations of rock strata, coloured in on a printed map by his friend and supporter, the map publisher, John Carey.

**A Delineation of the Strata of England
and Wales, 1815, William Smith**

Courtesy the British Library

William Smith's great geological map of England and Wales was first published, against enormous odds, in 1814. It was the last such work to be produced almost entirely based on the field work of a single individual. It was also the first to show an entirely different interpretation of England, that of the underlying rocks and strata of the country and, as such, challenged entrenched religious and scientific opinion. With this single work, Smith would pre-empt Darwin and much else in scientific thought in the following years.

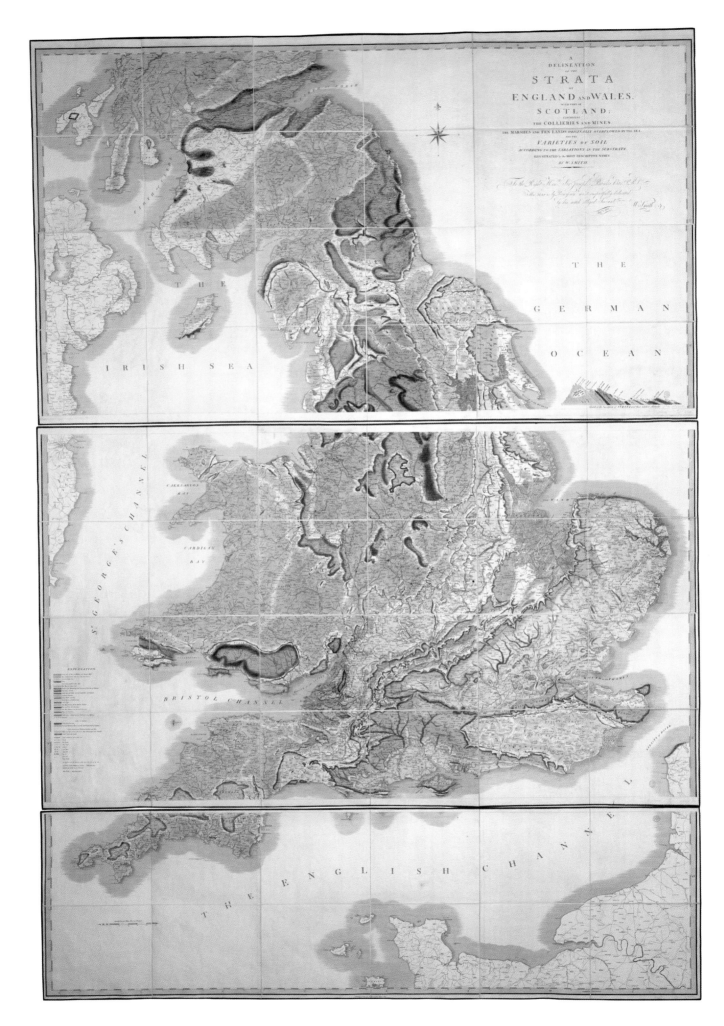

Geological Map of England, 1849

Courtesy Science Museum Pictorial/
Science & Society Picture Library

Just a few years after the publication of William
Smith's controversial map, A Delineation of the
Strata of England and Wales, the geological map of
England and Wales became a commonplace sight in
school classrooms. This poster from 1849 by John
Emslie (published by James Reynolds) is one of a
series of 44 scientific teaching diagrams for use in
lectures and display on classroom walls. It reflects
the Victorians' thirst for scientific knowledge of the
natural world, but also shows how the map had
become a normal component in communicating
complex ideas to a non-technical audience.

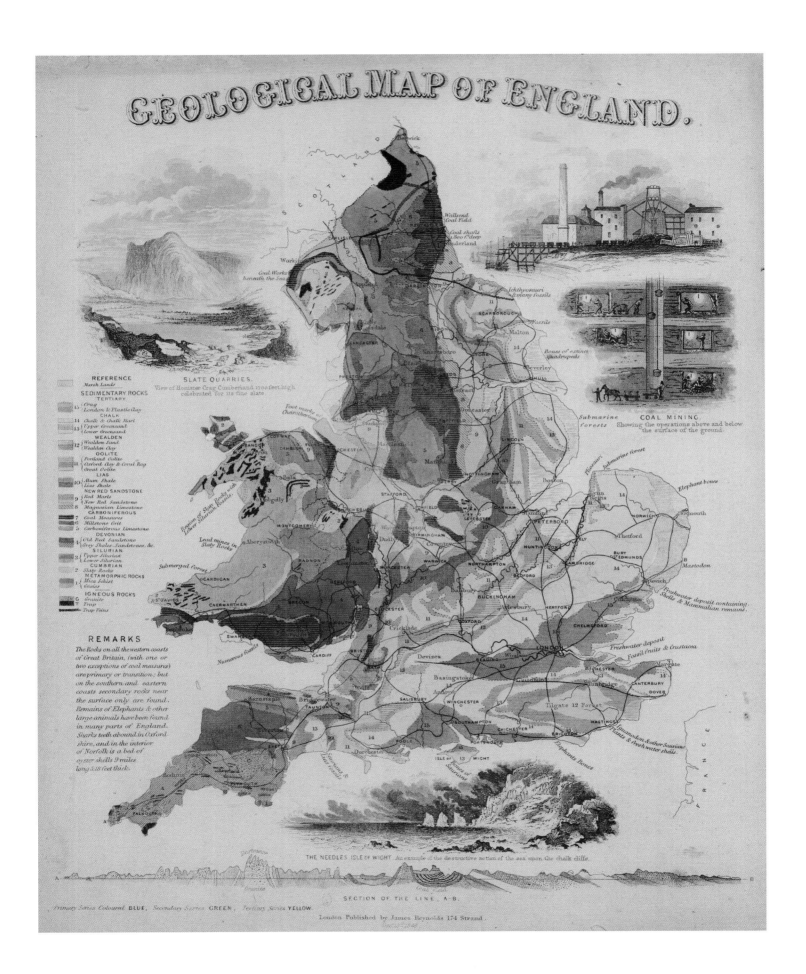

GEOLOGICAL MAP OF ENGLAND.

USING THE LAND

In parallel to the work on subterranean England, intensive work was being carried out to survey, draw and quantify developments on the surface. Advances in agriculture in the eighteenth century were changing the way the land was used. Landowners were determined to enclose fields, to protect and manage their crops, to 'improve' underproductive land and transform the Medieval farming system into a high-performing industry capable of growing more produce and—more importantly—of turning a tidy profit. Fields began to be enclosed as the result of Acts of Parliament from the beginning of the seventeenth century, but this only really gathered momentum after 1760, as WG Hoskins records in his classic work *The Making of the English Landscape*:[13]

> The total area dealt with before 1760 could hardly have exceeded 400,000 acres, a negligible amount when one thinks of England as a whole—only just over one per cent. In the next 40 years no fewer than 1,479 enclosure acts were passed, dealing with nearly two and a half million acres. Altogether, between 1761 and 1844, there were more than 2,500 acts dealing with rather more than four million acres of open fields.[14]

Enclosure, as a process, required maps. Indeed, they were a legal requirement under many of the acts. Sir John Sinclair's enclosure bill of 1797 explicitly spells out this principle: "The surveyor of surveyors shall, with all convenient speed, make an exact survey of all such common arable fields, common meadows, common pastures, wastes or commons."[15] As a result—despite having a relatively poor survival rate—enclosure maps are well stocked in local archives. The job of preparing them was a straightforward one and probably a welcome addition to the regular tasks undertaken by the local land surveyor (whether employed directly by the landowner or not). George Eliot, whose own father was a land surveyor, celebrates the occupation in *Middlemarch* (set during the period 1830–1832) in her description of that solid citizen, Caleb Garth: "Though he had only been a short time under a surveyor, and had been chiefly his own teacher, he knew more of land, building, and mining than most of the special men in the county."[16]

The result was relatively good mapping of the productive countryside, field by field, and at a large scale. Two classic examples are the accurate and detailed counterpart to John Rocque's map of Berkshire of 1761 and Thomas Milne's Plan of the Cities of London and Westminster, circumjacent towns and parishes: laid down from a trigonometrical survey taken in the years 1795–1799. Milne's map, in particular, shows detail of the land use in each enclosed field. (The trigonometrical survey referred to in the title is Roy's measured line on Hounslow Heath.)

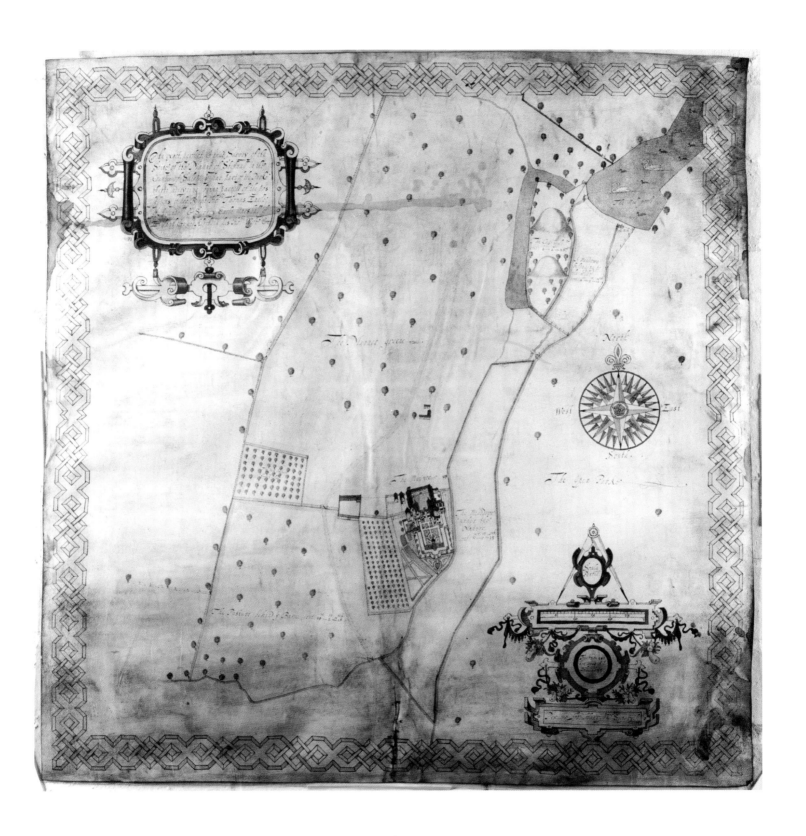

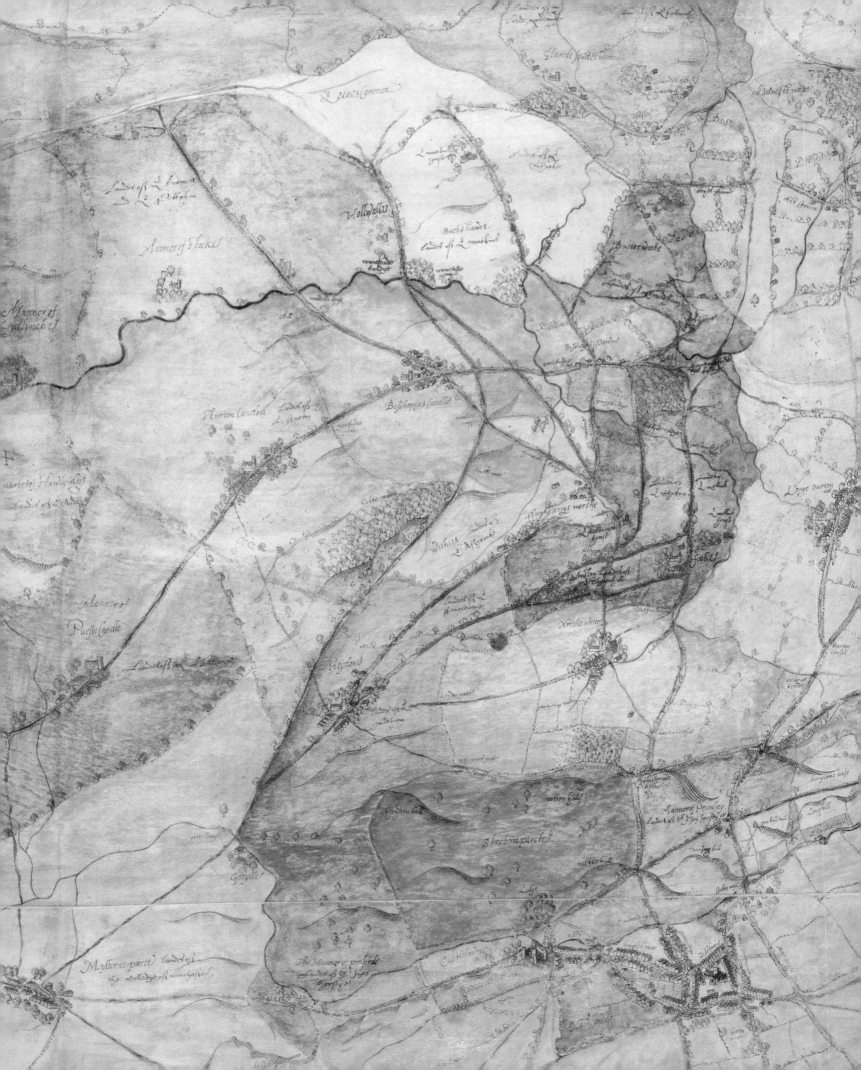

**Map of the Manor of West Harting, 1632,
William Cotham**

Courtesy the British Library

An estate map drawn for the Caryll family of their
Manor at West Harting on the South Downs. The
Manor was granted to Sir Edward Caryll by the
Crown in 1610 and his son John Caryll must have
been keen to record and improve the estate.

Enclosure mapping would prepare surveyors for the next land use challenge: the Tithe Commutation Act of 1836. The Act replaced the frequently oppressive and highly unpopular practice of giving the local clergy ten per cent of farmers' agricultural produce (as payment for the services the clergy offered) for a system that was notionally fairer but much more complicated. A sufficiently accurate cadastral (ownership) map of each parish was necessary to calculate the contribution of each farmer, whether in cash or kind. The question became one of how best to tackle this requirement. Robert Dawson, now the map adviser to the Tithe Commission (the body overseeing the implementation of the Act), believed it to be a unique opportunity to carry out a 'General Survey or Cadastre', an accurate nationwide property survey, similar to that achieved in many European countries in recent years. It would take far more time, effort and resources but the results would have been worth it. The country would have had accurate information on property for tax liabilities as well as a tool to assist in planning how to combat potential food shortages (as the country had suffered during the Napoleonic Wars).

Despite 11 per cent of the country being surveyed to the standard he proposed, Dawson would not see his ambitions achieved. Instead, local surveyors would produce tithe maps at different scales, to a multitude of standards and with varying degrees of accuracy. The opportunity to map England comprehensively and to record accurate details of land use and ownership had been lost. Like so many other grand schemes for nationwide endeavors it came unstuck. The Ordnance Survey had many decades of work yet before it would have comprehensive, large scale surveys and mapping in place and the first half of the nineteenth century was not a period when mapping England seemed politically worthwhile.

A thorough land use survey had to wait until the 1930s. Following the experience of the First World War, and with the threat of another on the horizon, the necessity for an official survey of the country was imperative. Led by the geographer, L Dudley Stamp, the First Land Utilisation Survey of Great Britain used the resources of school teachers and pupils to carry out the on-the-ground research, marking up 1:10,560 Ordnance Survey with information of seven different categories of land use. The results—printed on 170 sheets—were published between 1933 and 1946. The Survey was invaluable to wartime resource planning, however, its findings were almost immediately made invalid as all sorts of land was expropriated and ploughed up in order to feed a nation under partial naval blockade.

As access to mapping has become easier and more accessible in recent years—with aerial photography, satellite images, and particularly computer-based technology, providing maps with the opportunity to show all kinds of data potentially aligned with land use (tenure type, house cost, demographics, health and crime statistics)—we have seen

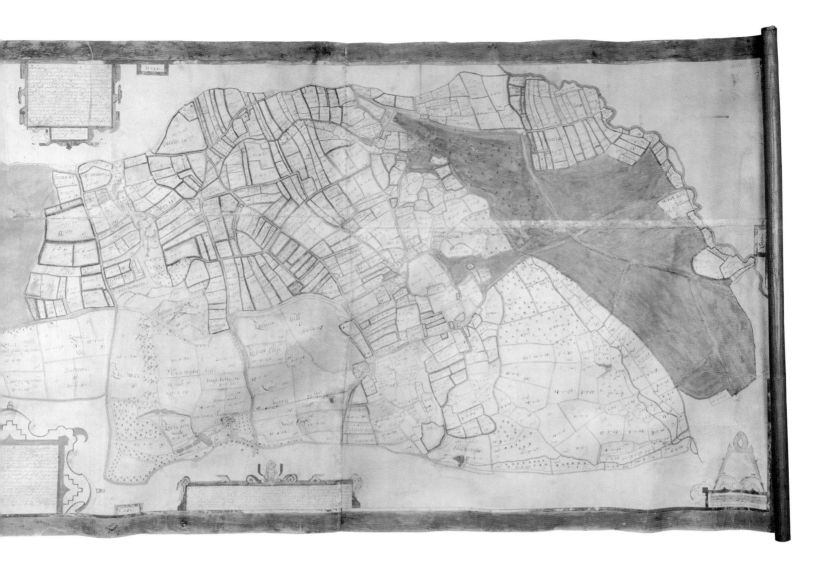

a proliferation of such maps, with, literally, more appearing each day. They may no longer threaten the country's 'identity' as such but they do continually and restlessly attempt to reveal its character whether it be through newspaper graphics, on television, or the internet and advertising. We may be less worried about new maps of our country, because there are now so many of them.

**Plan of Matthew Baillie's Estate
at Duntisbourne, Gloucestershire,
1790–1823, J & W Newton**

Courtesy Science Museum/
Science & Society Picture Library

The map of Matthew Baillie's (the famous anatomist)
estate in Gloucestershire illustrates a style of
rigorously functional estate mapmaking concerned
with the ownership, enclosure and use of the land.

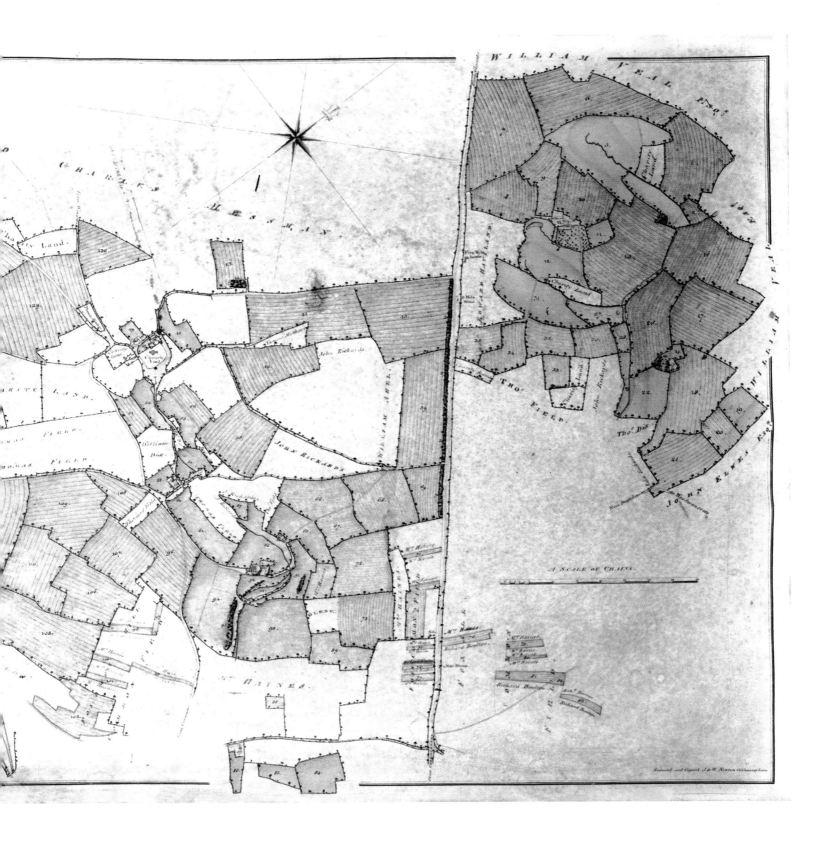

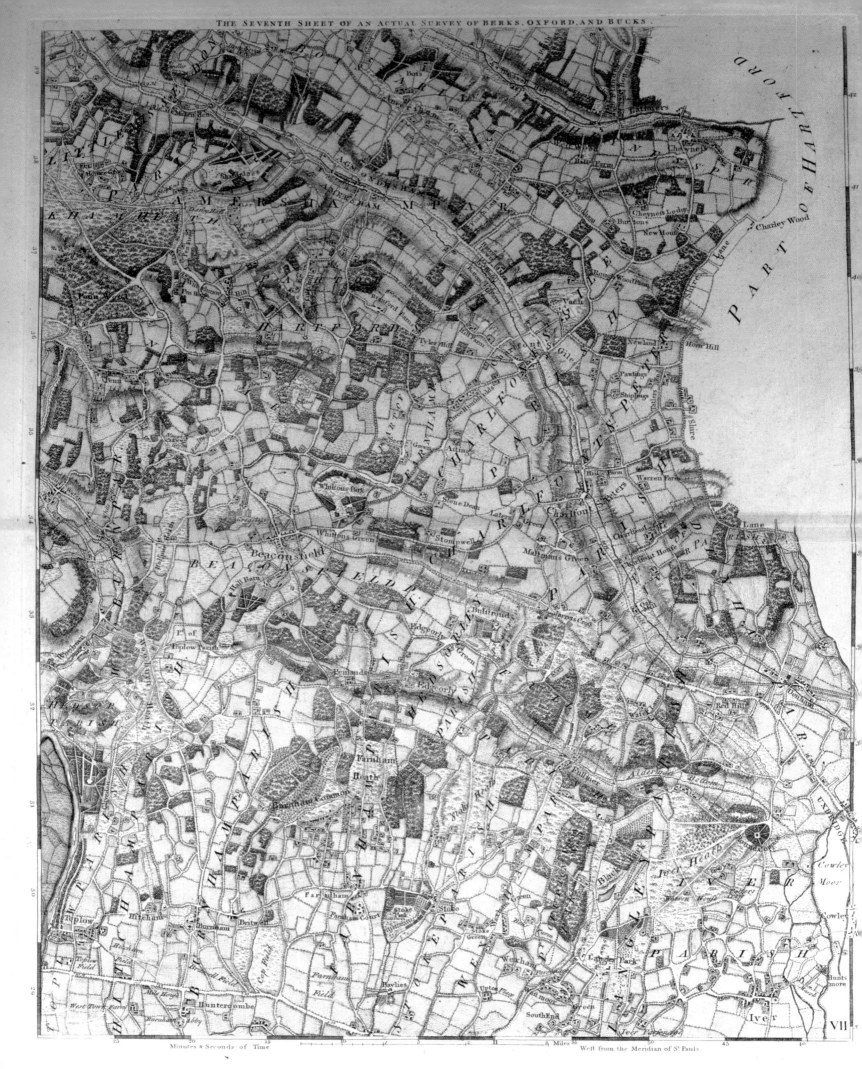

PART OF HARTFORD

PART OF HARTFORD

Charley Wood

Horn Hill

Beaconsfield

Farnham Heath

Farnham

Hitcham

Taplow

Burnham

Huntercombe

Stoke

Stoke Green

Iver

Cowley Meer

Cowley

VII

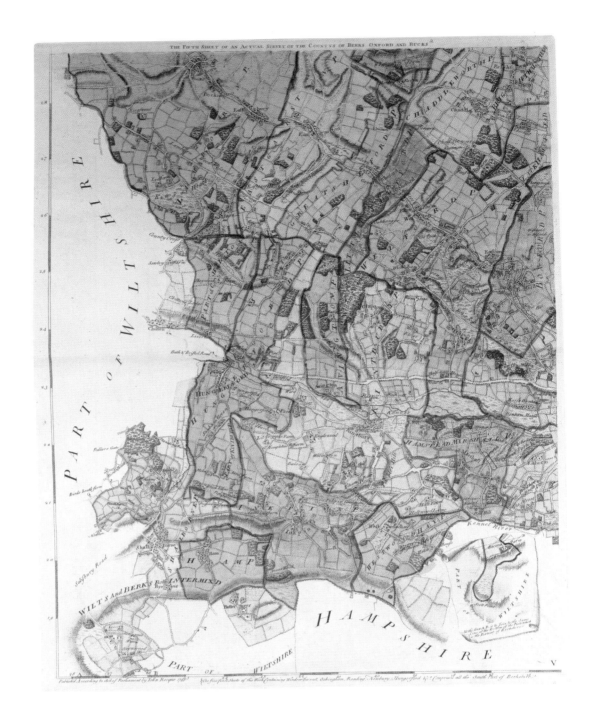

Opposite (detail) and above **County of Berkshire, John Rocque, 1750**

John Rocque—the former *dessinateur des jardins* and one of the most accomplished and stylish cartographers of the eighteenth century—was fascinated by land use and dwelt heavily on the detail of almost every field and copse (at least this is the impression given by his intensely beautiful maps). In practice much of this detail communicates something of the character and feel of topographical mapping even if its scale and accuracy are incorrect.

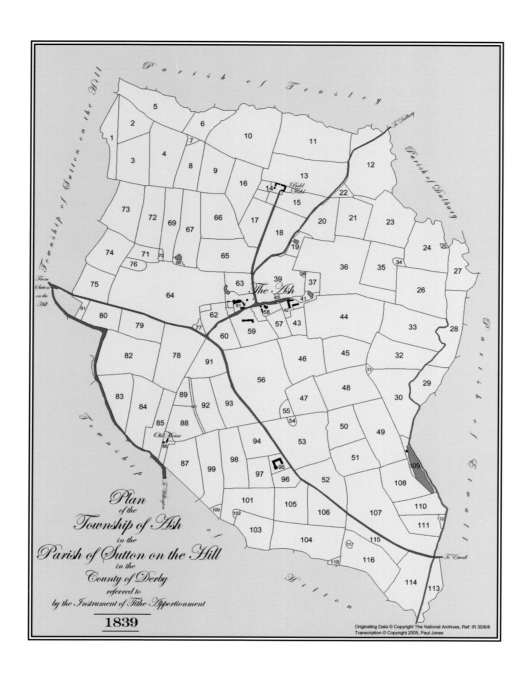

Plan
of the
Township of Ash
in the
Parish of Sutton on the Hill
in the
County of Derby
referred to
by the Instrument of Tithe Apportionment
1839

Originating Data © Copyright The National Archives, Ref: IR 30/8/8
Transcription © Copyright 2005, Paul Jones

Tithe Map and Apportionment of Thurvaston, Oslaston and Cropper in the Parish of Sutton on the Hill, Derbyshire, 1839

This Tithe Map shows the apportionment of the land in the parish to different individuals.

An accompanying schedule provides a brief description and the details of the occupiers, the owners, the type of land (old turf mowed, new turf pastured, wood, arable, meadow, etc.) and its area. From these designations, a value of annual tithe would be calculated. Three copies of these maps and schedules were prepared, one of which was held centrally by the tithe commissioners, one locally in the parish church and one in the diocesan registry.

Tithe Map of the Parish of Stoke, 1842

The first systematic and large scale survey of agricultural land in England and Wales was carried out for tithe mapping, following the Tithe Commutation Act of 1836. In contrast to the general uniformity of tithe apportionments, these maps vary greatly in accuracy, scale and size. Surveys and valuations included reports on the existing condition of the farms that, in the case of the Stoke and Stapleton Estate, were mostly neglected. These reports also included observations on improvements, proposals for remodelling the farms, introducing better methods of cultivation and the use of new chemicals to improve wheat production. This map, although a copy based on an earlier rendering, retains the field names as recorded in the 1842 Survey and Valuation for the Beaufort Estates.

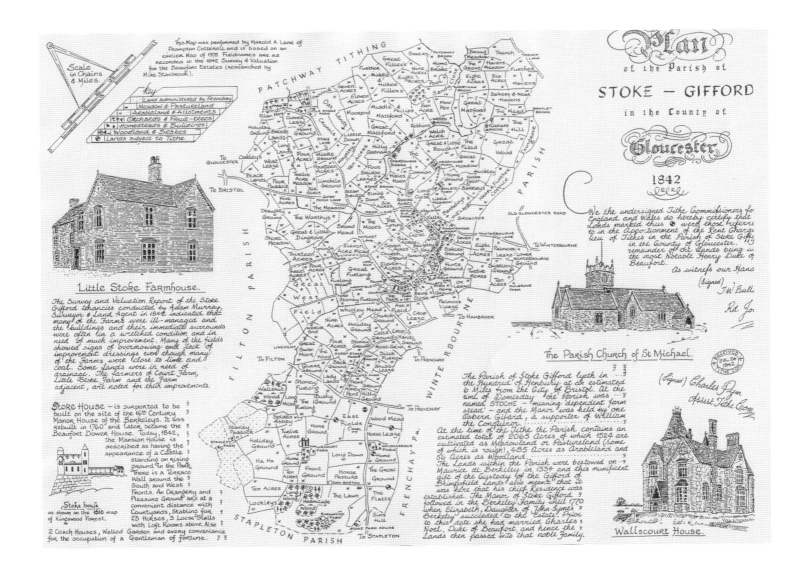

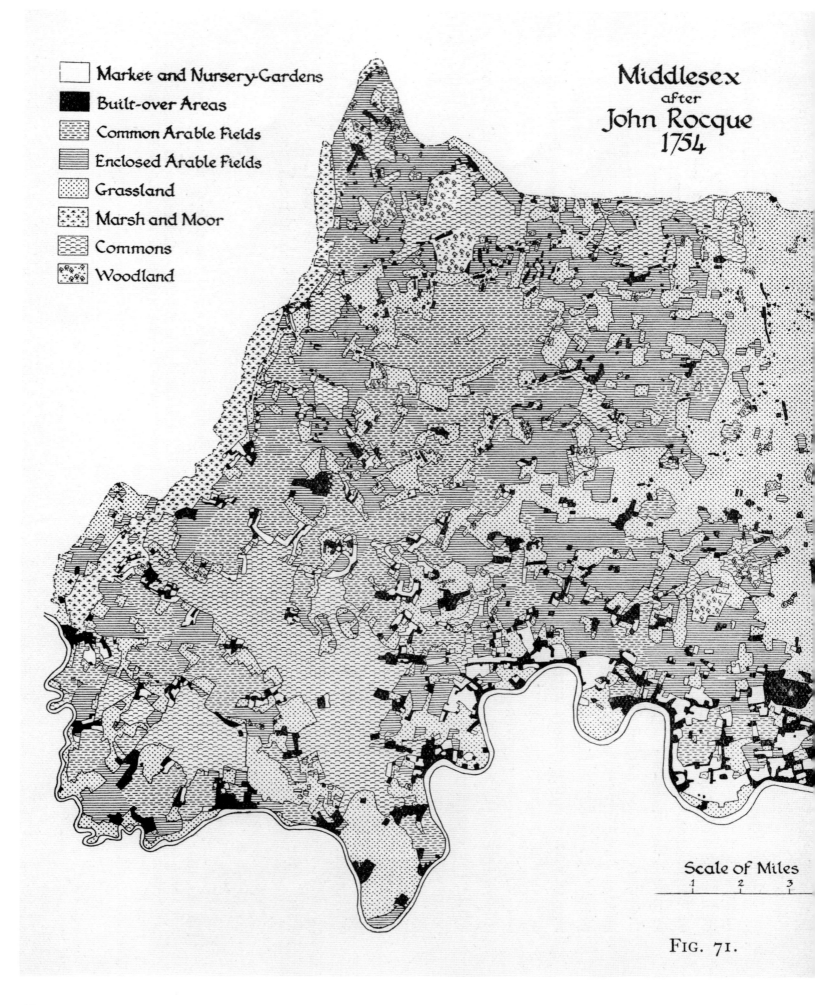

Market and Nursery-Gardens
Built-over Areas
Common Arable Fields
Enclosed Arable Fields
Grassland
Marsh and Moor
Commons
Woodland

Middlesex
after
John Rocque
1754

Scale of Miles
1   2   3

FIG. 71.

**Middlesex after John Rocque, 1754,
from the Second Land Utilisation
Survey of Great Britain, 1960s**

The Second Land Utilisation Survey of Britain was
undertaken in the 1960s by Alice Coleman of King's
College, London, and followed similar guidelines to
Dudley Stamp's initial survey of the 1930s. Over
30,000 volunteers contributed to the survey, which
resulted in Ordnance Survey plans overprinted
with graphics depicting a wide range of different
uses (from the residential and commercial to
detailed industrial and agricultural usage). Deploying
aerial photography rather than eyes on the street,
subsequent surveys have been far less labour
intensive (although the 1996 Land Use Britain survey
organised by the Geographical Association did return
to the original methodology by employing evidence
from over 50,000 school children).

**The Built Over Areas of Middlesex in 1935
from the Second Land Utilisation Survey
of Great Britain, 1960s**

Another example of a map from the Second Land
Utilisation Survey of Great Britain, this one depicts
the built-over areas of Middlesex in the 1930s.

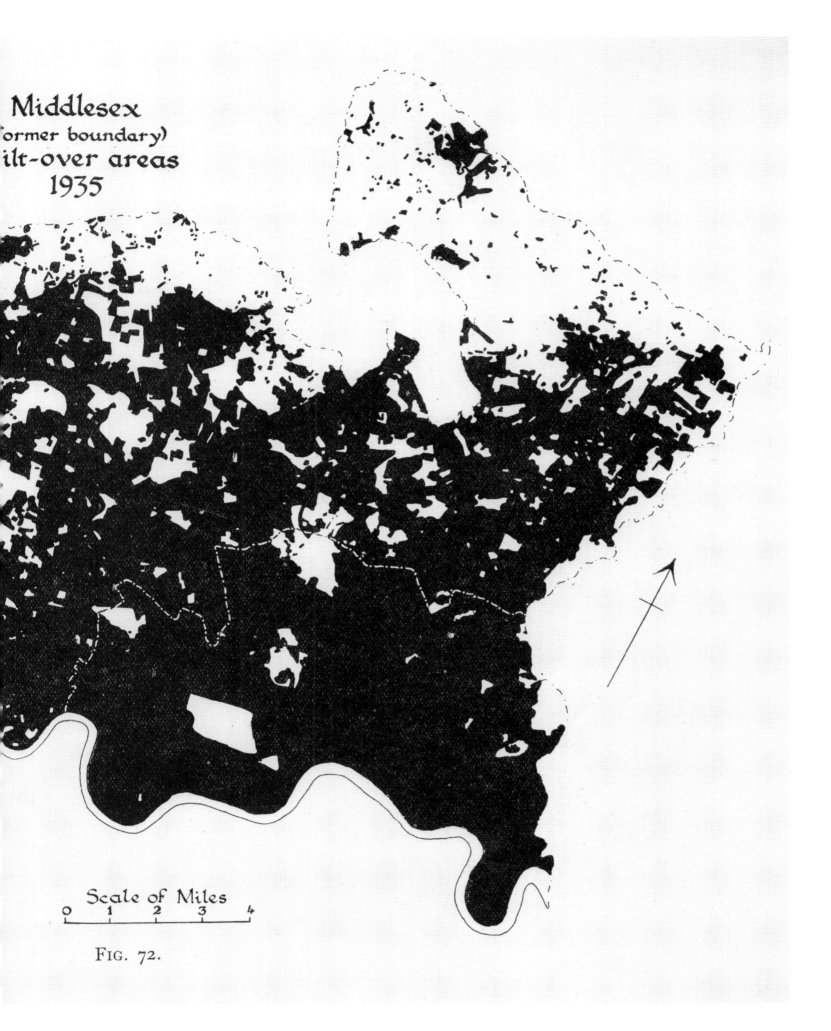

Middlesex
(former boundary)
built-over areas
1935

Scale of Miles
0 1 2 3 4

FIG. 72.

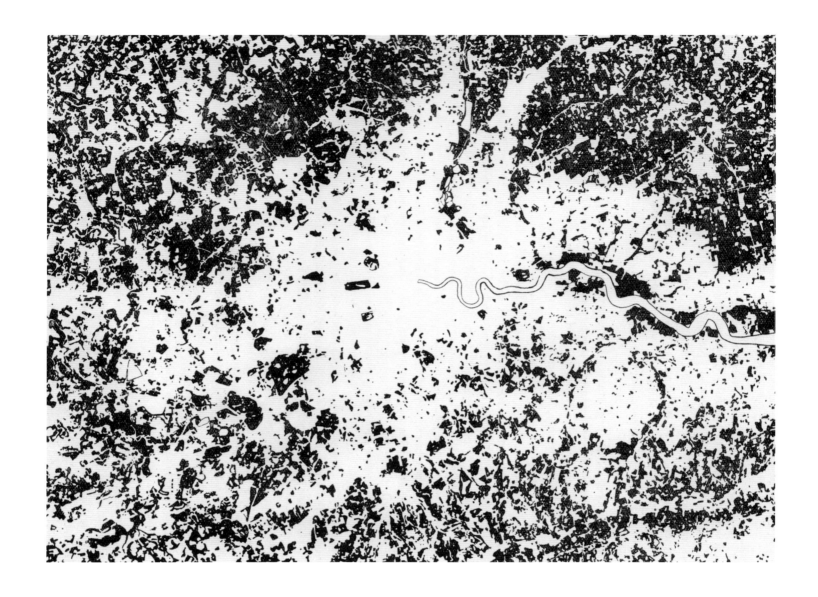

**Permanent Grass in London and its Environs
from the Second Land Utilisation Survey of
Great Britain, 1960s**

This land utilisation survey map depicts the location
of permanent grassy areas in and around London in
the 1960s.

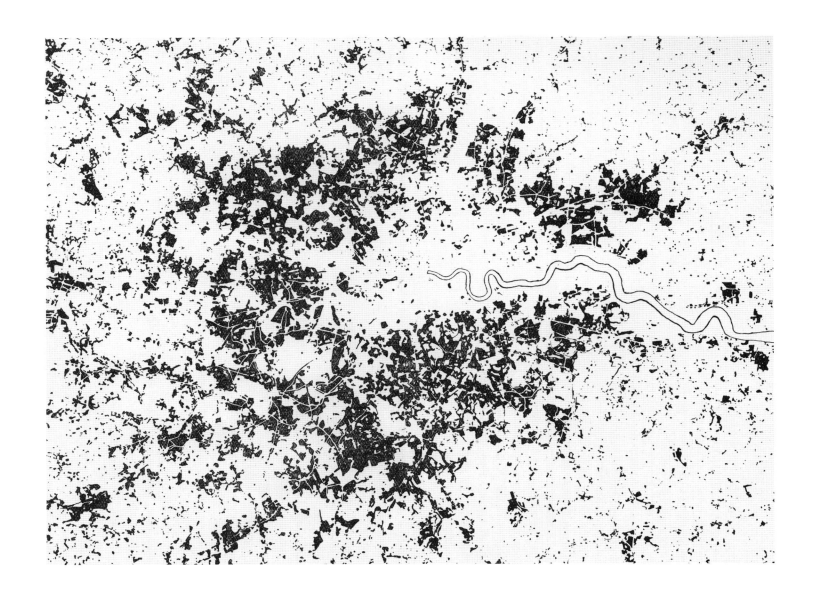

**Houses with Gardens in London and its
Environs from the Second Land Utilisation
Survey of Great Britain, 1960s**
The Second Land Utilisation Survey of Great
Britain investigated all manner of land use, including
this map of houses with gardens in London and its
outlying areas.

**Map of the British Isles elucidating
the Distribution of the Population based
on the Census of 1841 and 1849,
August Heinrich Petermann**

Courtesy the British Library

The Population Census of 1841 was the first modern
census of Britain and the model for all subsequent
population counts in the country. The census
counted every individual within each household,
providing the first detailed survey of the nation. This
groundbreaking map by the German cartographer,
August Heinrich Petermann, was developed soon
after and provides an instant overview of the nation's
urban and rural structures. Although such intensity
of technique would not be repeated in later mapping
(as contours and printed colour became prevalent
later in the century) this map influenced many future
such renderings and was used for both informative
and propaganda purposes.

Map of England

showing prevalence of Cholera, 1849

Courtesy the Wellcome Library, London

A map of England depicting the impact of cholera during
the epidemic of 1849. The relative degree of mortality
is expressed in the darkness of the shading. The dates
indicate the time at which the epidemic broke out.

Map of Leeds, 1842, from "Report to Her Majesty's Principal Secretary of State for the Home Department, from the Poor Law Commissioners, on an inquiry into the sanitary condition of the labouring population of Great Britain"

Courtesy the British Library

The great—if at times misguided and awkward—Victorian social reformer, Edwin Chadwick, produced this report (in collaboration with Dr Thomas Southwood-Smith) on sanitary conditions across Britain at his own expense. It was part of his, essentially effective, campaign to sting politicians into action and to address the open sewers and contaminated water supplies affecting the rapidly expanding city slums, whose infrastructures were failing to meet the exceptional demands being made of them. This map from his report shows the impoverished and unsanitary areas of Leeds.

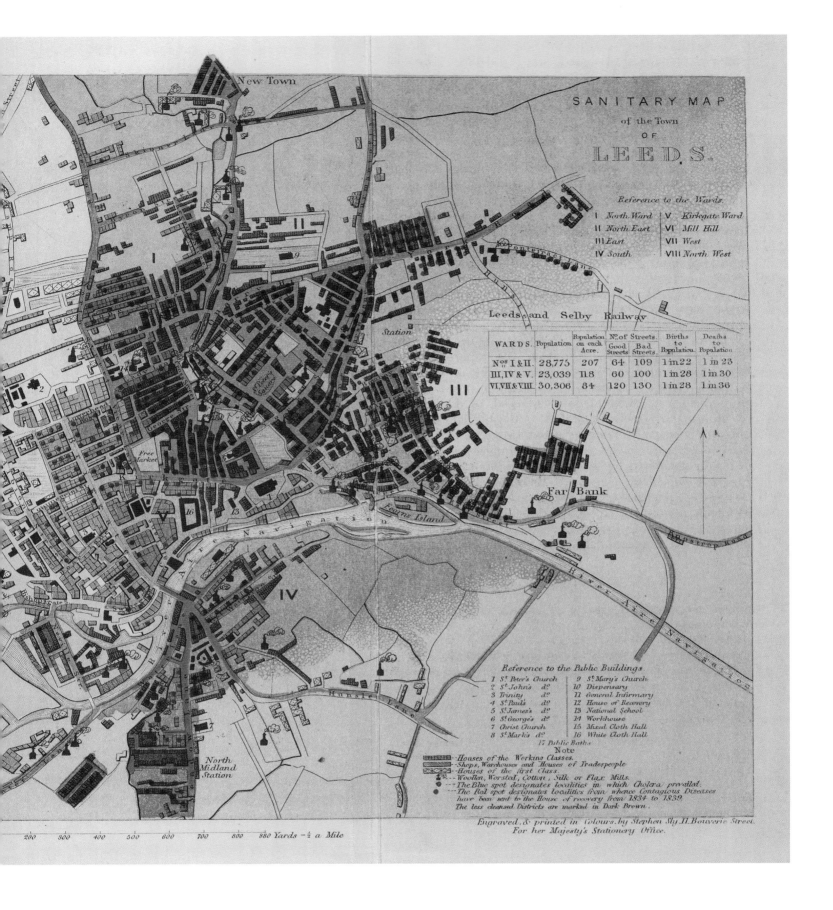

SANITARY MAP

of the Town

OF

LEEDS.

New Town

Reference to the Wards.

I	North Ward	V	Kirkgate Ward
II	North East	VI	Mill Hill
III	East	VII	West
IV	South	VIII	North West

Leeds and Selby Railway

Station

WARDS.	Population.	Population on each Acre.	No. of Streets.		Births to Population.	Deaths to Population.
			Good Streets.	Bad Streets.		
Nos I & II.	28,775	207	64	109	1 in 22	1 in 23
III, IV & V.	23,039	118	60	100	1 in 28	1 in 30
VI, VII & VIII.	30,306	84	120	130	1 in 28	1 in 36

III

Far Bank

Leeds Island

Free Market

River Aire Navigation

River Aire Navigation

IV

Hunslet Lane

North Midland Station

Reference to the Public Buildings

1	St. Peter's Church	9	St. Mary's Church
2	St. John's do.	10	Dispensary
3	Trinity do.	11	General Infirmary
4	St. Paul's do.	12	House of Recovery
5	St. James's do.	13	National School
6	St. George's do.	14	Workhouse
7	Christ Church	15	Mixed Cloth Hall
8	St. Mark's do.	16	White Cloth Hall
17 Public Baths			

Note

Houses of the Working Classes.
Shops, Warehouses and Houses of Tradespeople
Houses of the first Class.
Woollen, Worsted, Cotton, Silk or Flax Mills.
The Blue spot designates localities in which Cholera prevailed.
The Red spot designates localities from whence Contagious Diseases
have been sent to the House of recovery from 1834 to 1839.
The less cleansed Districts are marked in Dark Brown.

Engraved, & printed in Colours, by Stephen Sly, 11 Bouverie Street.
For her Majesty's Stationery Office.

200   300   400   500   600   700   800   880 Yards = ¼ a Mile

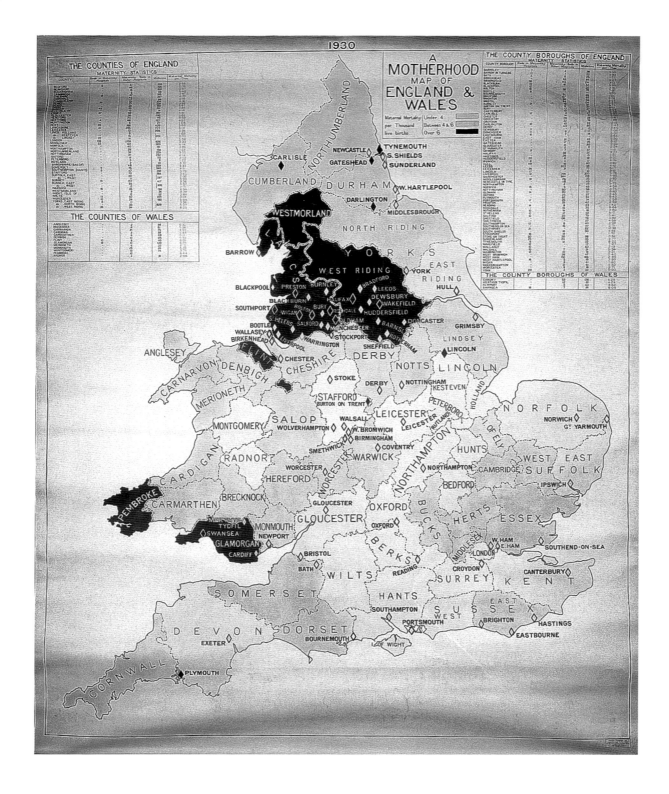

National Birthday Trust Fund Papers:

Motherhood Map, 1930

Courtesy the Wellcome Library, London

The National Birthday Trust Fund was established in 1928 to address the high incidence of maternal mortality in England and Wales. This Motherhood Map is the result of an early attempt to assess the scale and pattern of motherhood across the country.

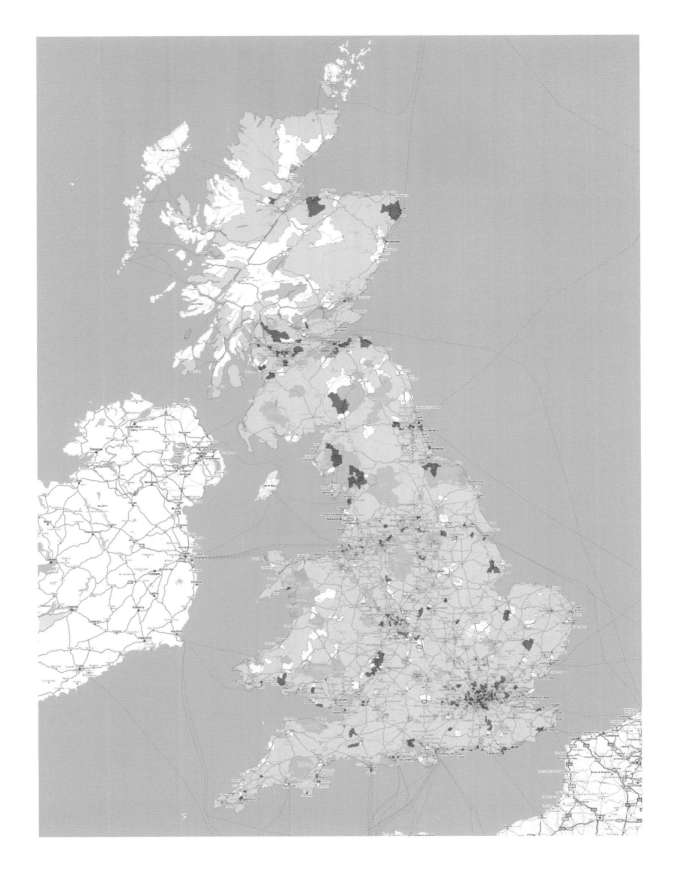

Mapping the Credit Crunch, 2008

Courtesy CASA, University College London

The Centre for Advanced Spatial Analysis at University College London is the source of much experimentation with the recombinant merging of maps and information. In the case of this interactive map produced in collaboration with BBC Radio's *Today* programme it sought to map the mood of the country in response to the impact of the credit crunch and its effect on property prices in 2008. This rendering has little scientific validity but does demonstrate the potential for using maps to capture the less tangible aspects of national life as they develop.

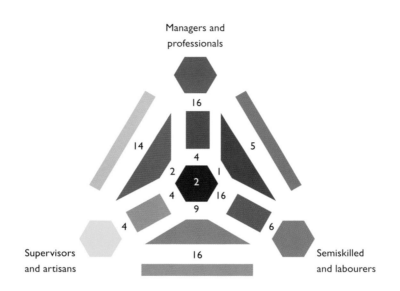

Managers and
professionals

16

14   5

4

2   1

2

4   16

9

4   6

16

Supervisors
and artisans

Semiskilled
and labourers

Only workers with
no occupation are
under-represented:

1

Figures are % of all residents in Britain.

Over and under-representation of social
groups above national average proportions
in categories nearest to labels.

Workers with no occupation are over-
represented in the seven inner categories.

Scale ☐ = 250,000 people

Social mixing from *A New Social Atlas of Britain*, Daniel Dorling/Wiley, 1995

Courtesy Daniel Dorling/Wiley

The integration of social data and cartography has been one of the relatively unsung achievements of the twentieth century, even though we find simple versions of such images on an almost daily basis in our newspapers. More complex versions, communicating sometimes highly information-rich data in a single image, have been developed by academics attempting to find effective ways of revealing patterns in otherwise forbidding columns of figures. Computers have, in turn, greatly assisted this effort and it is now possible to translate raw data—for example from a nationwide census—very rapidly into map images.

A New Social Atlas of Britain mainly uses 1991 national census data to produce page after page of map-diagrams showing everything from population to work patterns to car ownership, in a series of 'mappings'.

Freeze Logical ViewPoint
Freeze Displacement Map
Freeze Image Map
Show Clip Level Boundaries
Colour-Code Batches
Shader View
Highlight Overdraw
Wire Overlay
Enable All Levels
Disable Displacement Mapping
Fog
Show Map
Height Select
Alpha Up
Alpha Down
Shininess Up
Shininess Down
Show Axis
Wireframe
Top View
X View
Z View
Ground Align Eyes
Terrain Origin Position
Terrain Centre Position
Cycle Colours
Show Console
Speed:0.02mph

Opposite **Equal Population grid Squares, 1995, Daniel Dorling**

Courtesy of Daniel Dorling/Wiley

From August Petermann in 1849 on, mapmakers have been searching for ways of expressing population density in a clear way. Dorling's map, based on figures from the 1991 census, uses a grid methodology to give equal emphasis to the different areas of the country; each square containing 30,000 people; and in so doing creates a compelling map of Britain.

Above **Screenshot of Snowdon in Wales as visualised by GeoVisionary software, 2007**

Courtesy GeoVisionary Virtalis

GeoVisionary is a cutting-edge mapping system developed by Virtalis in collaboration with the British Geological Survey. Combining a powerful data engine with a virtual geological toolkit, GeoVisionary enables geoscientists to visualise, analyse and disseminate large datasets seamlessly in an immersive, real-time environment.

THE ENVIRONMENT

Viking, North Utsire, South Utsire, Forties, Cromarty, Forth, Tyne, Dogger, Fisher, German Bight, Humber, Thames, Dover, Wight, Portland, Plymouth, Biscay, Trafalgar, FitzRoy, Sole, Lundy, Fastnet, Irish Sea, Shannon, Rockall, Malin, Hebrides, Bailey, Fair Isle, Faeroes, Southeast Iceland.[17]

While the British Isles and the surrounding seas are the *locus operandi* of the shipping and weather forecasts, these quintessential, map-based descriptors evoke images of England that are more memorable than many other cartographic renderings (usually by way of our overexposure to them). The history of mapping the weather has a particularly strong connection to England; having emerged from the combination of deeply committed scientific enquiry and global exploration.

English explorers were a determined breed, committed to unraveling the mysteries—not only of the world—but the universe beyond. In 1724, Edmund Halley plotted the route of a full solar eclipse across the south of England, while, in 1768, Captain James Cook—along with the botanists Joseph Banks and Daniel Solanderset—sailed the Pacific on board his ship, the Endeavour, to observe the transit of Venus across the face of the sun. Between 1784 and 1785, Vincent Lunardi—the 'Daredevil aeronaut'—brought his hydrogen-filled balloon to England and Scotland only a year after the first manned ascent in the Montgolfier brothers' balloon in Paris.

It was the antics of one Robert FitzRoy, however, that would leave a lasting legacy on the charting of the environment. In 1831, FitzRoy was equipping his ship (HMS Beagle) in London to return to Tierra del Fuego, Chile. The previous year he had brought a group of the area's inhabitants to England, and was to take them home following their 'civilisation' in London. His friend and mentor at the Admiralty, Francis Beaufort, was instrumental in the organisation of the mission, not only enabling FitzRoy's command of the Beagle, but also in a finding a companion for the long voyage; a young scientist called Charles Darwin.

Whether the five years they spent sharing a shipboard cabin inspired FitzRoy's fascination with the natural world is impossible to say (although he did, towards the end of his life, become an ardent anti-evolutionist). FitzRoy would later become Meteorological Statist to the Board of Trade—not a significant position, but one from which he established a scientific collection of meteorological data, which would develop into a department of great importance at the modern Met Office.

His indefatiguable energies, when not expended on quarrelling with his enemies (even on occasion, challenging them to duels) were concentrated on issuing weather forecasts. His predictions, however, were generally rather unsuccessful.

What FitzRoy lacked in accuracy, however, he made up for by developing the first documented weather charts, and even invented the

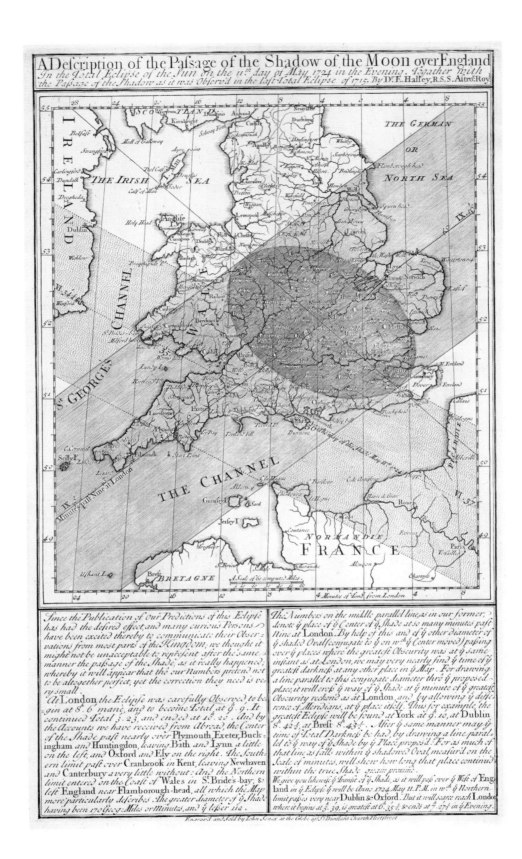

**Map and Description of the Eclipse,
11 May 1724, Edmond Halley**

Courtesy Science Museum Pictorial/
Science & Society Picture Library

This engraving by John Senex was published by Edmond Halley—the Astronomer Royal—and is an early example of mass observation of the land for scientific purposes. By collating the results of many reports he hoped to refine the ability to accurately predict such events as solar and lunar eclipses. (The results showed that the path illustrated was twice as wide as that observed on the day.)

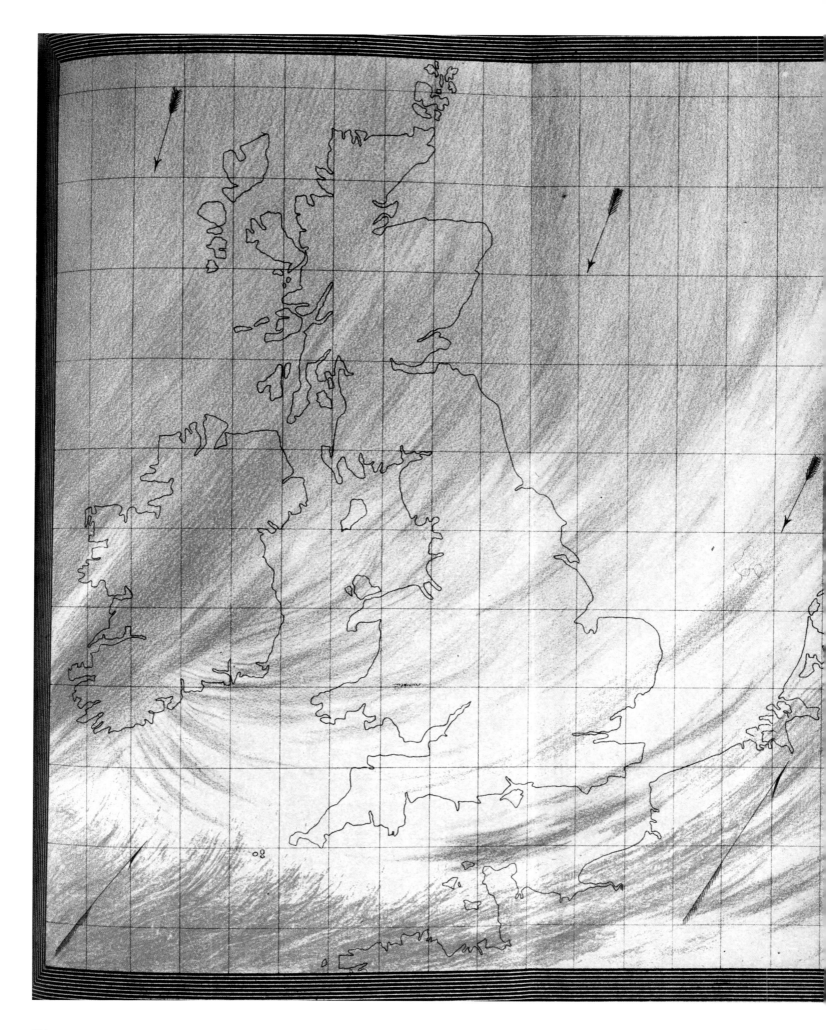

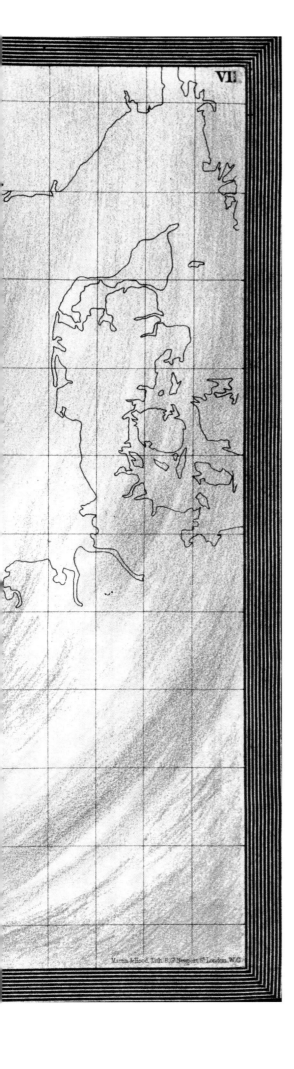

Atmospheric Disturbances, 1863,
from *The Weather Book*, Robert Fitzroy

Courtesy Science Museum Library/
Science & Society Picture Library
Robert Fitzroy, one-time captain of the HMS Beagle,
was Meteorological Statistician to the Board of
Trade when he developed the first weather charts of
England and, in many ways, established the most well-
known impression of the English in map form—as
a country obsessed with the weather. This example
from his *The Weather Book* shows the mixing of cold
polar air with warm tropical currents over both the
land and the North Sea.

WEATHER CHART, MARCH 31, 1875.

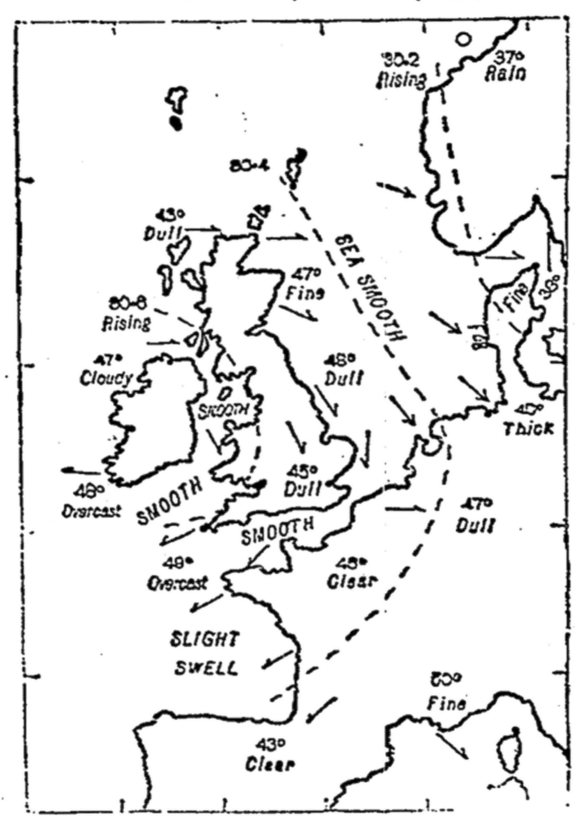

The dotted lines indicate the gradations of barometric pressure. The variations of the temperature are marked by figures, the state of the sea and sky by descriptive words, and the direction of the wind by arrows—barbed and feathered according to its force. ☉ denotes calm.

**First Newspaper Weather Chart, 1875,
as featured in *The Times***

In 1861 Sir Francis Galton, explorer, anthropologist, eugenicist, scientist, statistician and general polymath, published a paper in the *Philosophical Magazine* that proposed a means of producing a weather chart using movable type. On 1 April 1875 this became the first weather chart published in a daily newspaper, again produced by Galton. Now an everyday feature of such media Galton's work was a breakthrough in the use of maps to convey information about the country to the wider public.

term "forecasting the weather". England's abiding interest in the weather would have looked very different without him.

The weather chart—from its origins in naval practice and development through determined, if egotistical, intellects such as FitzRoy—has, in recent times, become the *de facto* means of communicating and analysing a much wider range of environmental factors, which now include pollutants and noise and other such elements of 'intrusion' in modern life. One of the most innovative methods of analysing today's environment can be found in Christian Nold's interpretative measurement of mood, emotion and levels of happiness across the country which, like the weather, is both variable and difficult to predict.

CLIMATE FUTURE

Mapping the climate has always had associations with forecasting the weather. However, while the implications of changes in the climate are less immediate and more predictable, they are far more significant. Long term trends can be mapped and projected in order to develop scenarios for climate change.

The main British climate change research facility, The Tyndall Centre for Climate Change Research, is named after a one time surveyor and draftsman in the Ordnance Survey. He later became a surveyor during the English railway boom in the mid-1840s before studying science under Robert Bunsen in Marburg, Germany, and going on to become one of Britain's foremost researchers and communicators of science. In the 1860s, he published research on the link between atmospheric composition and climatic variation and identified the 'greenhouse effect' as well as the importance of water vapour in the atmosphere: "This aqueous vapour is a blanket more necessary to the vegetable life of England than clothing is to man. Remove for a single summer night the aqueous vapour from the air that overspreads this country, and you would assuredly destroy every plant capable of being destroyed by a freezing temperature."[18]

The maps produced by the UK Climate Impact Programme to describe potential climate change in Britain not only provide a glimpse of England in the decades ahead—a changed country, with necessarily different patterns of land use reflecting a changing weather pattern, rising sea levels and a more precarious existence—they also reveal the ability of maps to carry narratives of the probable and the possible, to communicate something both of the past and the future with equal potency.

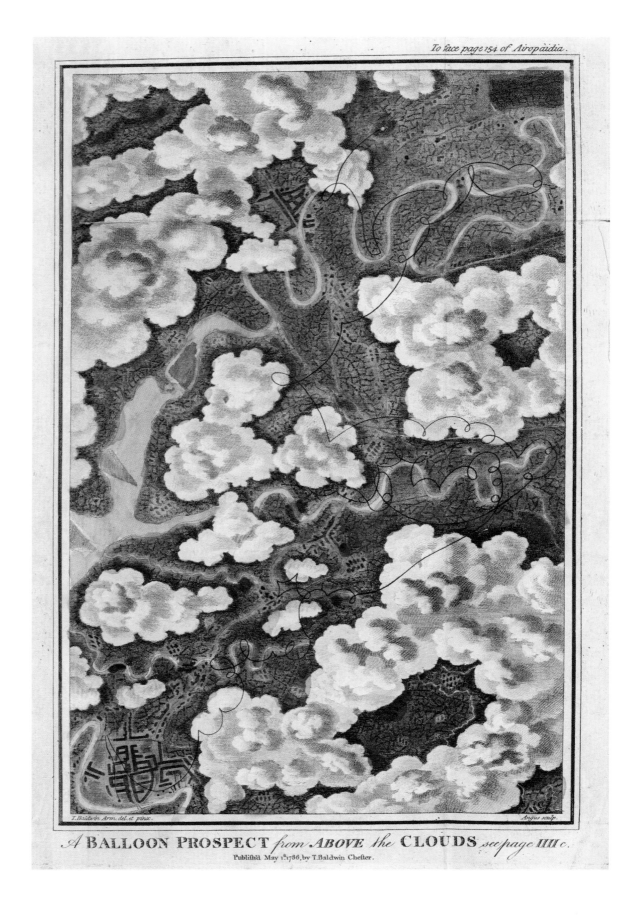

T.Baldwin Arm. del. et pinx.   *Angus sculp.*

A BALLOON PROSPECT *from* ABOVE *the* CLOUDS *see page IIII c.*

Publiſh'd May 1ſt 1786, by T.Baldwin Cheſter.

Above and opposite

**A Balloon Prospect from above the Clouds,
1785, Thomas Baldwin**

Courtesy Science Museum Library/
Science & Society Picture Library

The idea of viewing the country as a legible entity—
as it was depicted by maps of the time—became
a partial reality with advances in air travel. This
image is from *Airopaidia*, an illustrated narrative by a
young curate, Thomas Baldwin, of his solo flight (in a

balloon borrowed from the famous aeronaut Vincent
Lunardi) from Chester Castle in 1785. Baldwin's
imagery presages a later generation's use of aerial
images as a key source for mapmaking.

To be placed on Page 155 of Airopaidia.

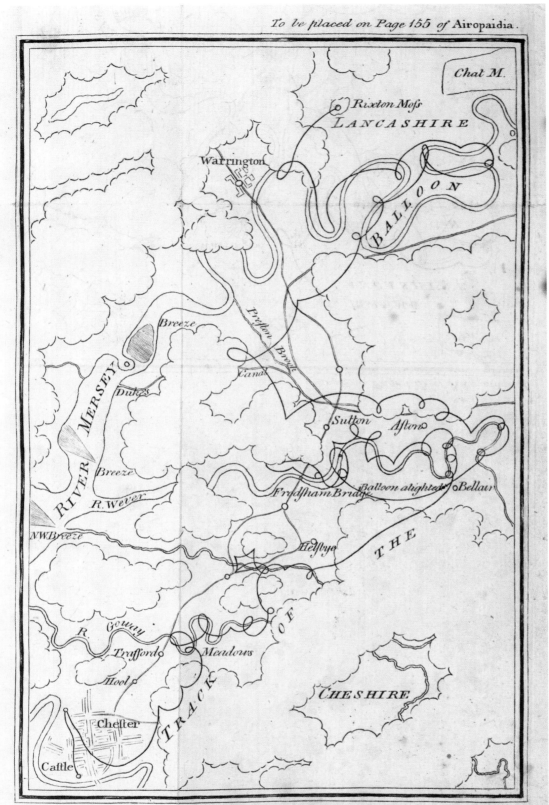

The EXPLANATORY Print.   see Page 1111 d.

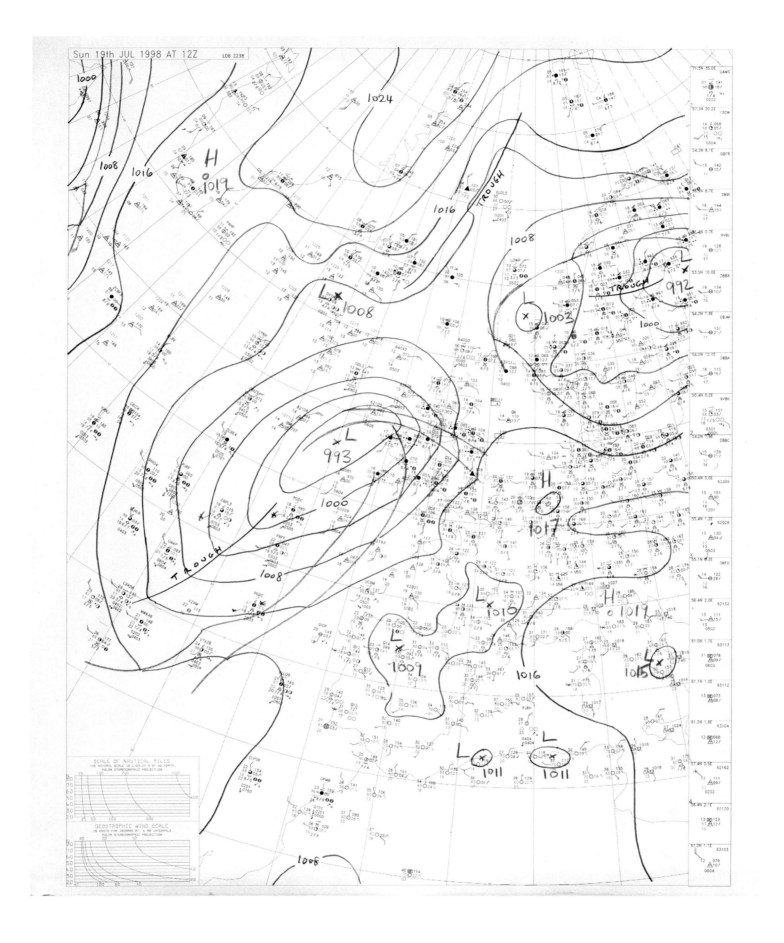

Sun 19th JUL 1998 AT 12Z   LDB 2238

THE INDEPENDENT

(Ireland, €1) 70p
Tuesday 26 June 2007
www.independent.co.uk
NUMBER 6,456

Your photographs

IN EXTRA

THE GREAT FLOOD

Torrential rain sweeps Britain in day of drama and devastation

Tynemouth
Up to three inches of rain, and near gale force winds caused rough seas.

Hessle
A man in his twenties died after his foot was trapped in a metal grating.

Leeds
More than 70 houses in the Halton area of Leeds had to be abandoned.

Wellington
A tornado swept across the countryside, reflecting the 'huge amount of energy' in the thunder clouds, according to the Met Office.

Cheltenham
Fire service inundated with reports of flood water breaching cellars; locals report roads like rivers.

Lydney
Up to 50 children were rescued after a bus became stranded in flood water in the Forest of Dean town.

Bideford
Heavy rain caused streams to overflow their banks, flooding dozens of properties.

Hull
Received at least four inches of rain; hundreds of homes flooded.

Sheffield
13-year-old boy missing after being swept away; traffic gridlocked as drivers abandon vehicles; at least 100 people rescued in Brightside area.

Wimbledon
Ten matches cancelled and many others, including Tim Henman's halted by rain.

Glastonbury
Thousands of fans delayed leaving the music festival; police handed out 3,000 space blankets.

Full report, page 2

Opposite **Weather Chart, 19 July 1998,**
Alan Howard

This is the last hand-drawn chart—superimposed on top of a standard template—printed from the main Meteorology Office database using the HORACE system, and produced at the London Weather Centre. After this map, rendering the production of weather charts would be fully computerised.

"The Great Flood", *The Independent,*
26 June 2007

Courtesy *The Independent*

The floods of 2007 and their impact on a number of English towns and cities create a strong graphic image for *The Independent* on the 26 June 2007. The combination of the weather, even in an extreme form and with tragic storylines, along with the image of England, make an ideal combination to appeal to the potential reader.

Key

Areas disturbed by noise & visual intrusion*

Undisturbed areas

Urban areas

* Areas disturbed by urban development, major infrastructure projects and other noise and visual intrusion.

N
W E
S
0   25   50 Km

Above (1960s) and opposite (2007)

CPRE Intrusion Map of England, 1960s and 2007

Courtesy CPRE

In 2007, as part of a long-standing battle to preserve the English countryside, the Campaign to Protect Rural England produced maps showing the extent of both aural and visual intrusion in the 1960s, 1990s and 2007. The maps clearly depict the increase in

'intrusion' but might equally be interpreted to show greater accessibility or economic activity across the country. By the twenty-first century, maps became a standard part of the armory of many such campaigning organisations.

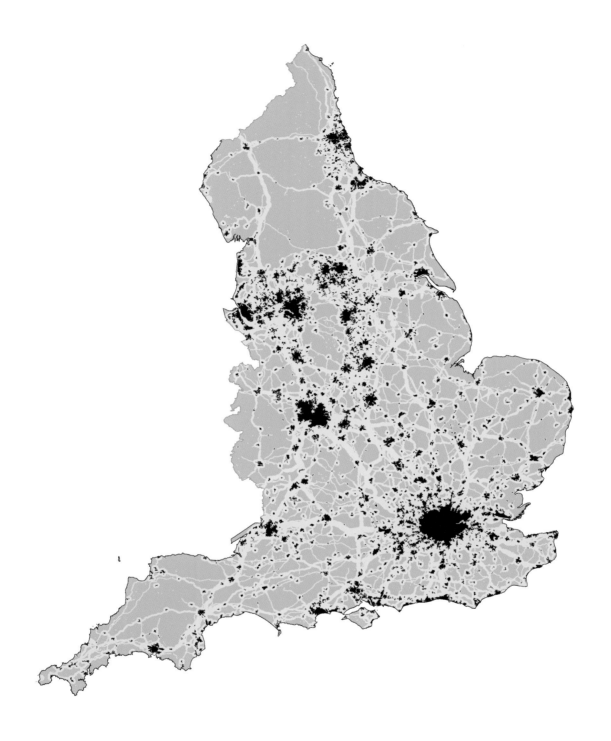

Defra Noise Map of England and Wales, 2008

Courtesy DEFRA/*The Guardian*

This 2008 Noise Map produced by the Department
for Food and Rural Affairs (DEFRA) deployed a
computer model of England rather than the ground
decibel readings to calculate noise levels. This map
assumes almost all noise is produced from modes
of transport—both on the road and in the air—but
does allow for the absorption of such noise by the
ground and buildings.

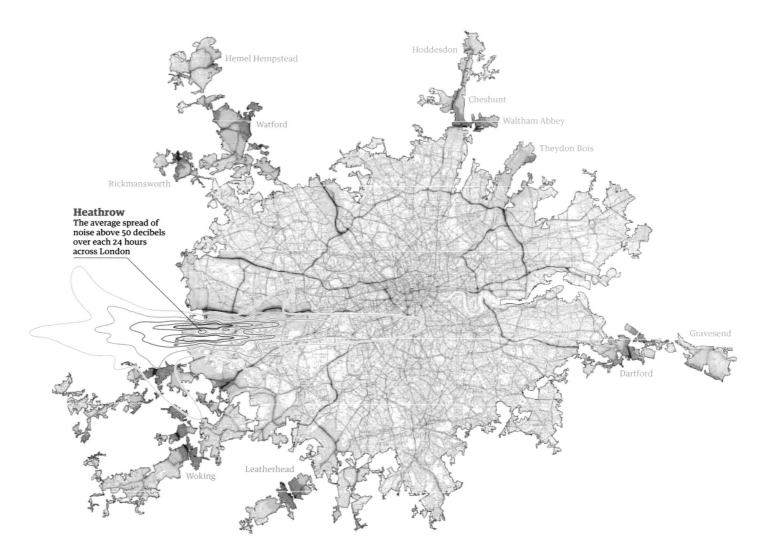

Heathrow
The average spread of
noise above 50 decibels
over each 24 hours
across London

Wolverhampton

Dudley

Birmingham

Newcastle

Gosforth

Washington

Chester-le-Street

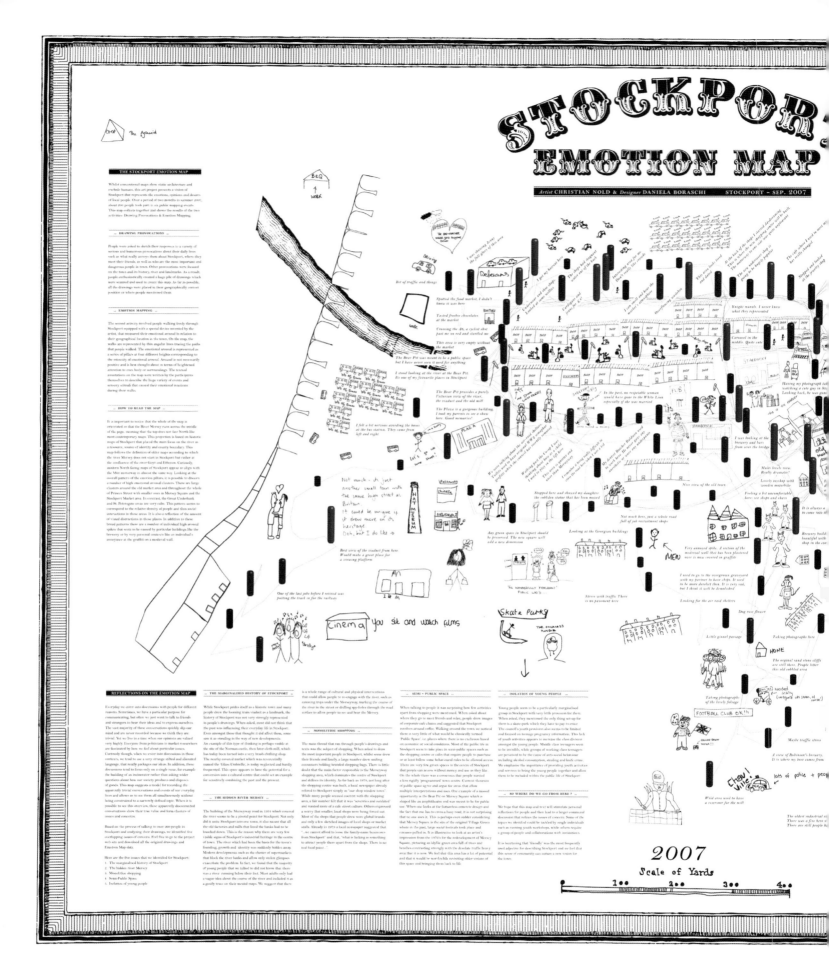

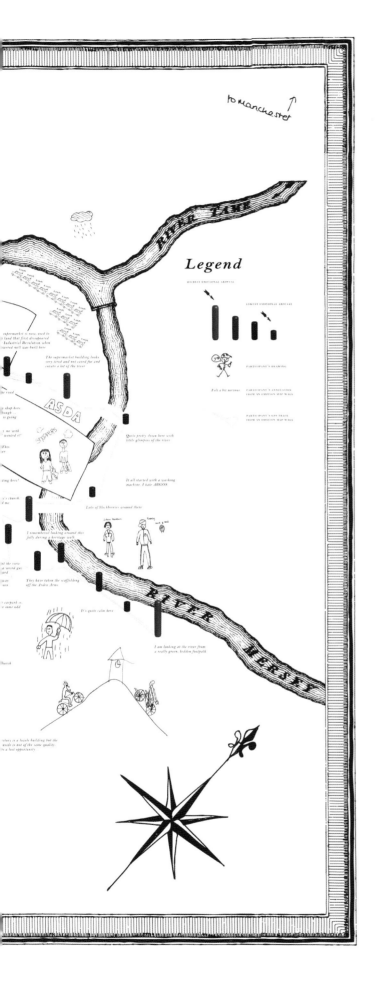

Stockport Emotion Map, 2007, Christian Nold

Courtesy Christian Nold

As mapping becomes more interactive greater links can be made between place and other factors—even emotion levels—as in this map of Stockport by the artist Christian Nold. Many of Nold's other maps utilise computer imagery to record such information, however, for this rendering (with its narrative incident and high levels of subjectivity) the artist has employed a more primitive method of representation.

TRANSPORT

In travelling through England, a luxuriance of objects presents itself to our view. Wherever we come, and which way so ever we look, we see something new, something significant, something well worth the traveller's stay.[19]

England is easily traversable; large proportions of travellers have been criss-crossing the country without too much effort since prehistoric times. The literary journey of England has been a staple for generations of writers including Daniel Defoe, William Cobbett, JB Priestley and Ian Nairn, to name just a few. As such, it has provided the opportunity to decry the loss and despoliation of the landscape and to search for a country that is still preserved in the popular imagination but which, like Lewis Carroll's imaginary land form "Hunting of the Snark", is far too elusive to trace.[20] However straightforward such journeys might seem, they would invariably have been much easier if the protagonists undertaking them had used a map.

Travel maps of England have been in existence since at least Roman times. Like many of the types of maps discussed in this book, however, there are extensive gaps between surveys, meaning that many a traveller would have relied on centuries-old information for long periods of time before they were updated.

The greatest surviving Roman map is the *Tabula Peutingeriana* (named after Konrad Peutinger to whom it was bequeathed in 1508 after its discovery in a library in Worms, Germany), a thirteenth century parchment copy of a Roman (probably fourth century) original. Drawn upon a narrow scroll of 6.75 metres in length, it depicts the *cursus publicus*; the routes used by the Imperial messenger service to send information rapidly from one part of the Empire to another. The *cursus publicus* reaches from modern Spain in the west to India and China in the east. Unfortunately, the westernmost leaf of the *Tabula* (which includes most of Britain) is missing, although the south coast up to Camulodunum (Colchester) is visible. In 1887, Conrad Miller reconstructed the map to include its missing parts.

While the original Peutinger map is inaccurate—being closer to Beck's London underground diagram in this sense—it remained the most widely used travel map across England, and many nations beyond, for several more centuries. This would only change with the arrival of the Gough Map in the mid-fourteenth century but, even then, the itineraries and maps derived from them were to dominate for many more centuries to come.

Despite the marked roads and routes depicted on the Gough Map (many of which did not necessarily exist) the most common method of navigation of England was to proceed from one village or town through to another using an itinery; a practice that would continue until as late as 1707. Such itineraries were distributed by hand; the Medieval equivalent of frequent flyers. The surviving examples of these maps can be found scribbled in the margins of other documents and, more frequently, in documents originating from monastery libraries. These itineraries would invariably describe the routes from mother foundations to satellite monasteries as well as give directions to reach the Holy Land itself. Matthew Paris, a monk of the Abbey of St Albans—as well as a prolific historian, illustrator and cartographer—transformed such an itinerary into a visual guide when, in the thirteenth century he recorded the route from London to Jerusalem. In so doing, Paris heralded the invention a new type of map; one in which the road leads towards the next town or landmark on to the eventual destination. This is a type of mapping still familiar to us today, as evidenced by every in-car satellite navigation device with their heads-up display and a prioritising of destination over orientation.

Paris left no cartographical disciples and, after this breakthrough in navigational mapping, roads would disappear from maps—with the mysterious exception of the Gough Map—for almost 500 years. Saxton, Speed and their Tudor contemporaries only depict bridges in their renderings, not the roads that might connect them. This is surprising in light of the fact that the main backer of such maps, William Cecil, had a keen interest in the state of roads and maintained records on their condition and the cost of upkeep. Sketches of Queen Elizabeth I's journey routes across England survive but, although map-like in their graphic simplicity, they have no other detail than the interconnections and distances between stopover points. It was distance that preoccupied seventeenth century cartographers, who would employ elaborate tables and diagrams that showed the number of miles between a multitude of places in their published works. The culmination of such mapping is John Adams' astonishing 12 sheet map *Anglia Totius Tubula cum Distantiis Notioribus In Itinerantium Usum Accommodata* of 1677. In Adams' work there are no marked roads but, rather, webs of spider-like lines each marked by its length that connect approximately 12,000 different locations. This map, despite its limited topographical information, was very popular and, as Delano-Smith and Kain note: "Several times reprinted, it was also reproduced as a smaller, more manageable, two sheet map. In fact, so successful was Adams' map that at least 18 imitative or derivative maps had been printed by 1750."[21]

Just two years before Adams produced his map, the country—despite its apparent obsession with distance and only distance—was provided with an alternative view of the journeys along its roads and highways. The mapmaker John Ogilby, bouncing back from the double disaster of having both lost his fortune in the English Civil War and then his entire publishing business and stock in the Great Fire of London, 1666, would produce his most accomplished work in 1675. Titled *Britannia, Volume the First or Illustration of the Kingdom of England and Dominion of Wales*, it harked back to the itinerary maps of Matthew Paris. *Britannia* includes a hundred plates delineating journeys across England to and from its larger towns and cities.

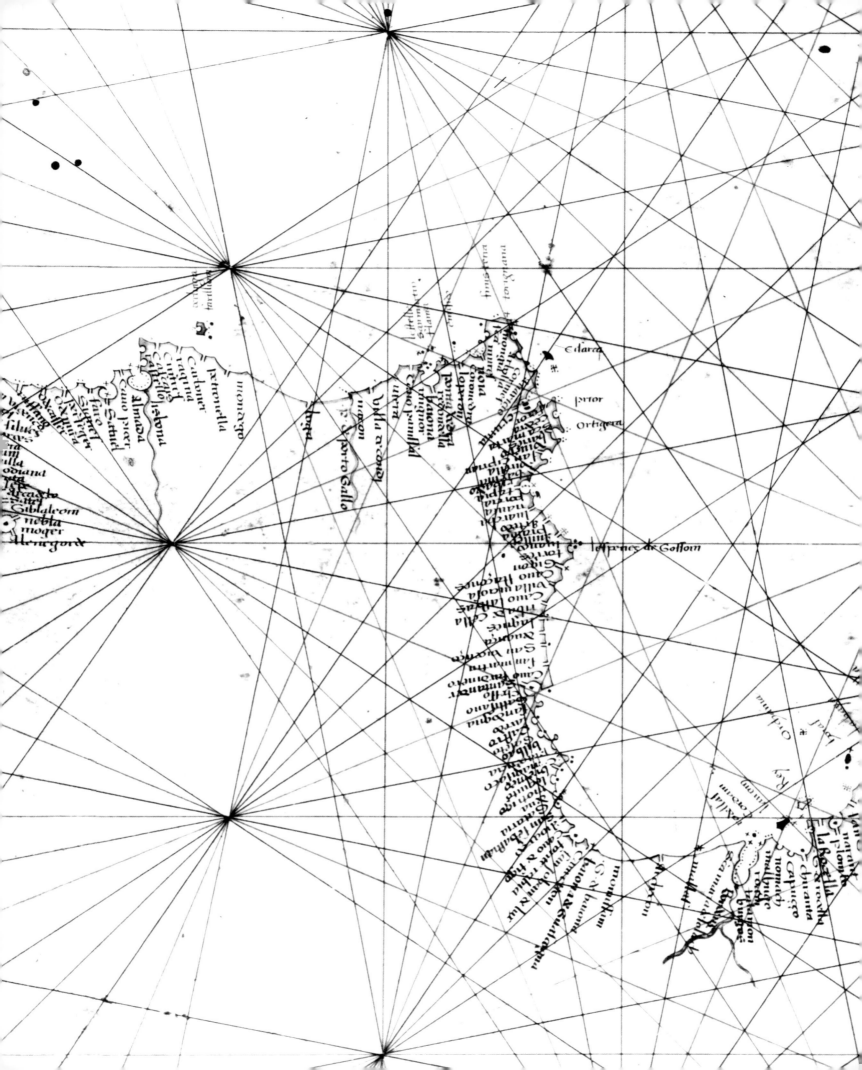

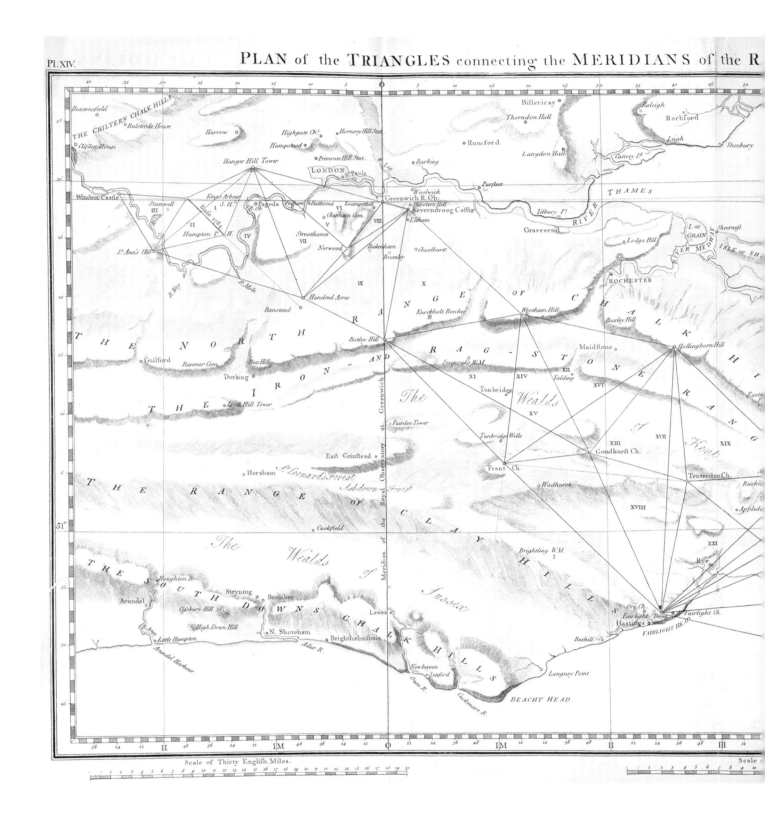

Scale of Thirty English Miles.

Scale

Plan of the Principal Triangulation of Britain, 1787–1790

Courtesy Science & Society Picture Library

The rivalry between French and English astronomers was of long standing by the end of the eighteenth century, with both sides claiming superiority for their observations, accuracy and instruments, as well as precedence for the 0 degree line of longitude running through the rival observatories in Paris and Greenwich. There was no doubt at this time that the French—under the direction of the Cassini family of astronomers, mathematicians and surveyors—had the upper hand in cartography. Cassini de Thury, third of the celebrated Cassini family of astronomers, Director of the Observatory in Paris and 'father' of French mapmaking, challenged the English to jointly measure the relative positions of the two observatories and to test the existing English calculation of the distance between them.

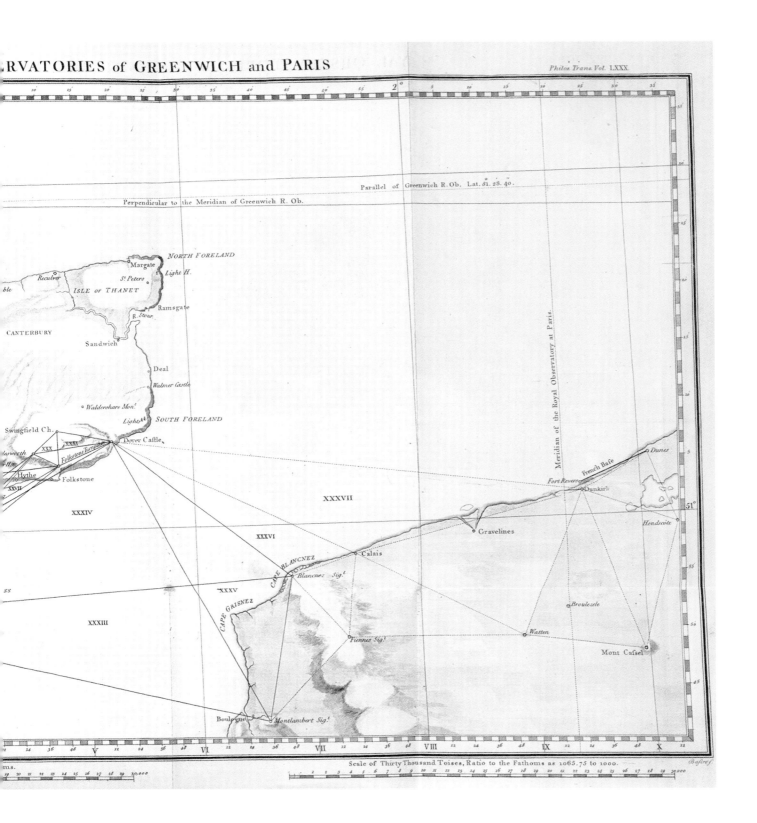

Philos. Trans. Vol. LXXX.

Parallel of Greenwich R. Ob. Lat. 51. 28. 40.

Perpendicular to the Meridian of Greenwich R. Ob.

NORTH FORELAND

Margate

Light H.

Reculver

St Peters

ISLE of THANET

Ramsgate

R. Stour

CANTERBURY

Sandwich

Deal

Walmer Castle

Walderehare Mon.

Swingfield Ch.

Lights

SOUTH FORELAND

Dasworth

XXXI

XXX

Dover Castle

H.

Folkstone Turnpike

Hythe

Folkstone

XXVII

XXXIV

XXXVII

French Base

Dunes

Fort Reverse

Dunkirk

XXXVI

Gravelines

Hondscote

CAPE BLANCNEZ

Calais

Blancnez Sig.t

XXXV

Meridian of the Royal Observatory at Paris

51°

CAPE GRAISNEZ

Broulezele

XXXIII

Fiennes Sig.t

Watten

Mont Cafsel

Boulogne

Montlambert Sig.t

Bafire f.

V   VI   VII   VIII   IX   X

ms.

10,000   20,000

20

The time of high water, and the motion of the Tide round the Coast of ENGLAND &c. Explaind

According to the Diurnal course of the *Moon*, her Rays make an Impression upon the *Æther*; and by its mediation upon our *Atmosphere*, and this also upon the *Sea* it self, whose Tides are so regularly known to follow $\tilde{y}$ Motion of the Moon, that it is constantly found by Observation that it is *High water* towards the *Pole* [*Example*] in our *Northern* Latitude of England, upon the Meridian when the Moon is also upon the *same Line* at the *Equator*, and by consequence the *Ocean* must be most deprefsed there, to raise the *Tide* here by refult from that deprefsion as its proper Cause.

To give the motion of the *Tide* a succefsive Defignation, firft as to *Place* whence it is to be drawn, it is the Center of the *Atlantick Ocean*, where it is rais'd by Libration and swells again by *Refult* from the new and Full Moons prefsure; but it is not full Sea there till they have attain'd to their Meridional height between the *Tropicks*; and so cannot reach *Ireland* by *Undulation* till after their Southing: hence the Primary Full-Moon, or Secundary new-Moon-tide makes at midnight and usually vifsets the West of *Ireland* at III, and the *Southern* Coaft at IV, in the morning, and moving towards the S.W. point of *England*, it swells near the *Lizard* about V, and pafseth by *Plymouth* at VI, *Dartmouth* at VII, *Portland* and *Weymouth* at VIII, by the *Isle of Wight* at IX, till it reach *Dover* at X½ pafseth the ftraight and leaving *Calais* at XI it reacheth the Chaps of Thames at XII (bu it is III before it getts to *London*) and at XII also comes to *Dunkirk*, pafseth by *Goree* at I½, and before the *Meas* at II½ and reacheth the Clifts of the *Texel* at III¼ ftill following the direction of the same fhoar all along the German Ocean.

But deflecting thence toward the S. Sea which lyeth before *Amfterdam* thro' the ftreights of the *Texel* at VI, and of the *Fly* at VII, and along the coaft of *Friezland* till VIII, when they conjoin into one at IX, and pafsing by *HORN* at XII, comes to *Amfterdam* at III, and arrives after all at *Harlem* where it finisheth its Course at IX.

To return to the Weftern coaft of *Ireland* whence the *Northward bound* division of the aforesaid Full and new Moon flood departed at the same time (vizt III, as the Eaftern flood) it arrives at the Weftside of the *Orcades* at IX in the morning, but being to encompafs all the *Shutland Islands*, and also to fwell the *whole extent* of Sea between the coaft of *Norway* and those *Islands* before the Flood could be reflected thence and reach the same height on the Eaft side of the *Orcades* it arrives not thither till III in the afternoon; nor did it affect the Road before *Fair Isle* which lies betwixt them till about XI, and it is obferv'd that it exceeds not XI½ along the Coaft of *Norway* till it hath pafsed the *Naize* and so continues to make at XII along *Denmark*, and the *Iutland Islands* till it reach the *Elbe* at the Eaft end of the *Dogger Bank*, where it meets the South Flood at the same hour, which meeting of the 2 Floods is the cause and making of the said Bank.

But as the *Tide* makes not on the Shoar of *Farr Isle* till about III in the afternoon; that is four Hours later later than it doth in the *Road*: so upon the Coaft of England and Scotland it continues to make so much later than on the opposite Shores of *Norway* and *Denmark*: i.e. either at or after III on the *Forelands* of both *Nations* till it reach *Flamborough Head* about IV, *Spurnhead* at V¼ and Lynn Deeps about VI. Thus so *regularly* Progrefsive is the Tidal Flood that its way is known even to an *Hourly* performance.

Chart showing the Sea Coast of England and Wales, Plate from *A Compleat Sett of Mapps of England and Wales in General*, 1724, Thomas Badeslade
Courtesy Yale Center for British Art/
Paul Mellon Collection/Bridgeman Art Library

As a surveyor/engineer, Thomas Badeslade, like many of his contemporaries, was fascinated by harbours and river schemes. This map, however, was a one off—reproduced from a manuscript atlas of ink and watercolour drawings on parchment bound into a sumptuous tooled and gilded leather book.

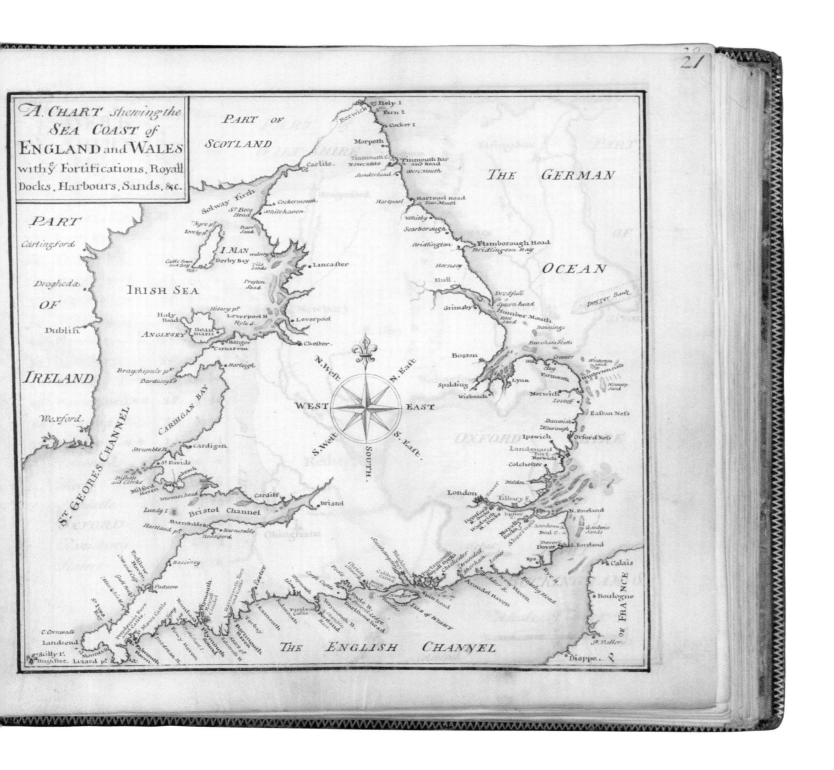

A CHART shewing the
SEA COAST of
ENGLAND and WALES
with ye Fortifications, Royall
Docks, Harbours, Sands, &c.

PART OF
SCOTLAND

PART
Carlingford
OF
Drogheda
Dublin
OF
IRELAND
Wexford

IRISH SEA

THE GERMAN

OCEAN

DOGGER Bank

ST GEORGES CHANNEL

CARDIGAN BAY

Bristol Channel

THE ENGLISH CHANNEL

OF FRANCE

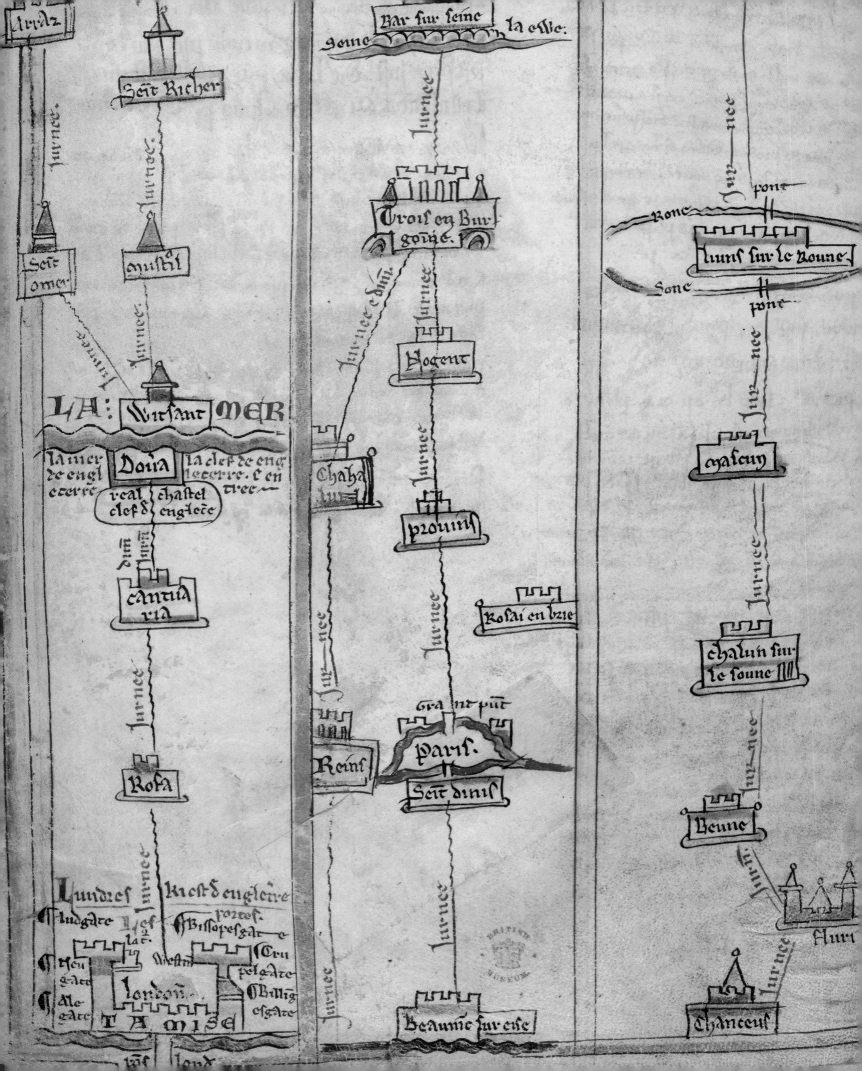

Each journey is drawn rigorously to scale, and leads the reader across and between hills, over rivers and through towns and villages. All twists and turns are presented in great detail and a graphic compass appears on each of the seven short lengths on each individual sheet. Distances are still shown but the map's key function is to relate something of the quality of the journey. (On a preparatory plate, all England's roads are depicted as they radiate out of London, an aesthetic convention that would not be out of place on many a modern road map of the country.) Ogilby wrote in the advertisement for the first volume of *Britannia* that:

> Having in our General Survey of all England design'd the actual admeasurement of above 40,000 miles of roads, and in order thereunto, already run over near two thirds of that quantity; we have in the subsequent work selected only the most considerable of them, or such as an orderly distribution of the kingdom has oblig'd us to exhibit.[22]

This is likely to be a considerable exaggeration and, most probably, an attempt to emphasise the level of work he had to undertake to bring such a groundbreaking project to fruition. He also made clear in the obsequious dedication "to his most serene and sacred majesty the high and mighty Prince Charles II" that he was attempting "to improve our commerce and correspondency at home by registering and illustrating your majesty's high-ways, directly and transversely, as from shore to shore and to the prescribed limits of the circumambient ocean from this great emporium and prime centre of the Kingdom, Your Royal Metropolis".[23] This was not to be some decorative object for the armchair traveller, it was intended for a very practical commercial object: making money, not only for Ogilby, but also for the nation as a whole.

Ogilby planned two subsequent volumes to complete *Britannia*: "The second, a description of the 25 cities with peculiar charts to each of them, but more particularly those of London and Westminster, and the third, a topographical description of the whole Kingdom."[24] They would have been astonishing volumes that might have redefined England, but Ogilby died only months later, leaving his flourishing mapmaking business to his step-grandson, William Morgan. (Morgan would have some success mapping London but did not attempt to complete the two outstanding volumes that Ogilby had left undone.)

If Ogilby's efforts were much plagiarised in the years following the publication of *Britannia*, his atlas of maps remained in print until the 1760s—including pocket editions—from 1719 onwards. Updated editions by mapmakers such as Daniel Paterson and John Carey also dominated the closing years of the eighteenth century. Accessible, easy to understand and concisely detailed, they must have made travelling seem both achievable and desirable.

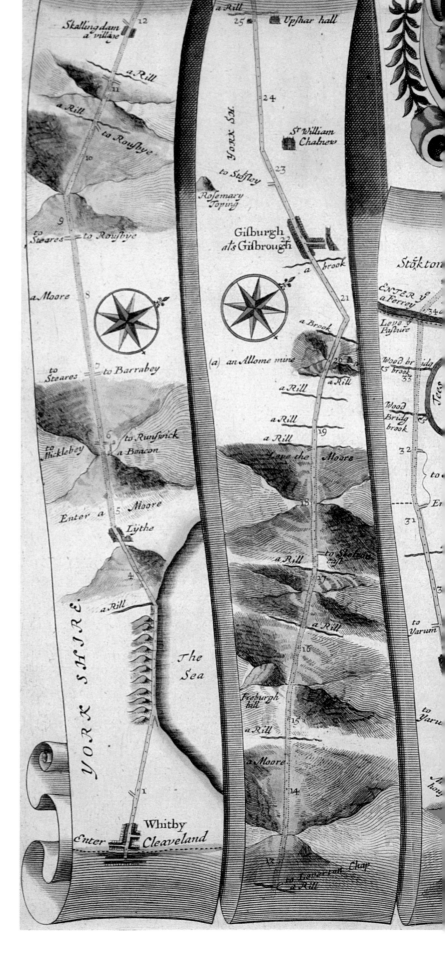

Road from Whitby to Durham, from John Ogilby's *Britannia*, 1675, John Ogilby

This strip road map transports the traveller through Whitby, over hills (rendered in John Ogilby's new and distinctive style), through villages, across streams and the River Tees, passing fields and woods of various descriptions, to Durham and then Tinmouth. Ogilby has provided plenty of diversion—albeit few alternative routes—along the way.

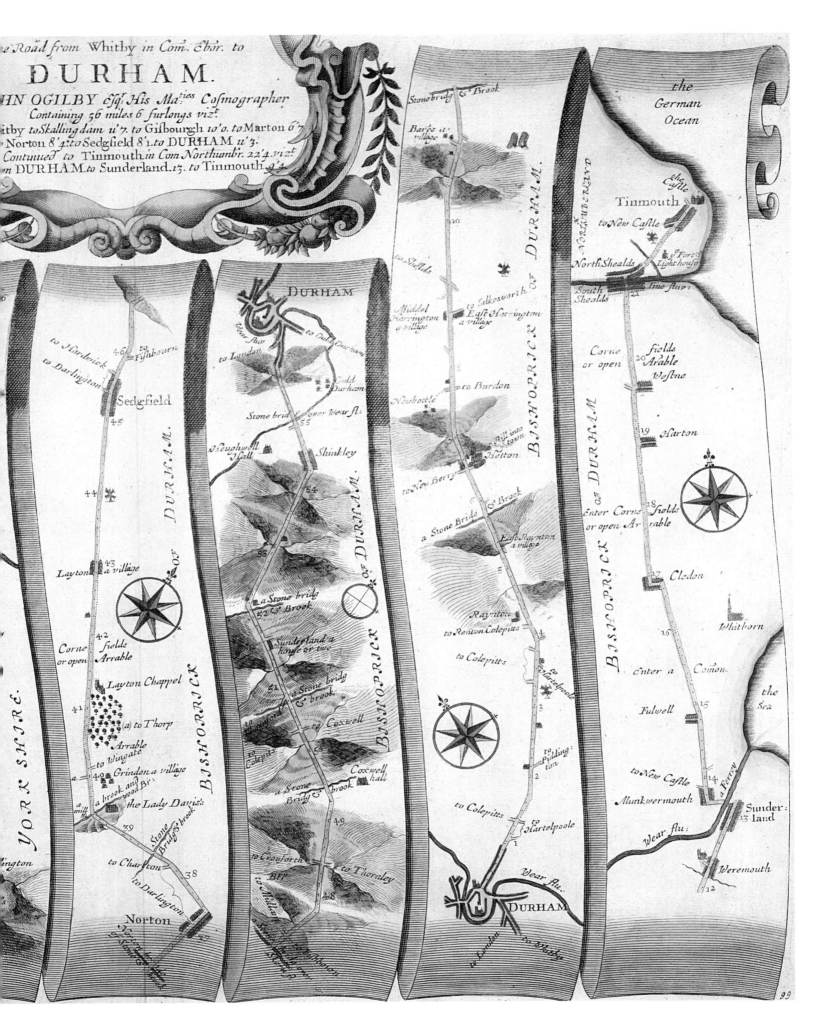

e Road from Whitby in Com. Ebor. to

DURHAM.

HN OGILBY Esqr. His Maties Cosmographer

Containing 56 miles 6 furlongs vizt.

itby to Skallingdam 11'7. to Gisbourgh 10'0. to Marton 6'7.

Norton 8'4 to Sedgfield 8'1 to DURHAM 11'3.

Contuued to Tinmouth in Com Northumbr. 22'4 vizt.

m DURHAM to Sunderland. 13. to Tinmouth. 9'4

the German Ocean

USEFUL AND INFORMATIVE

181

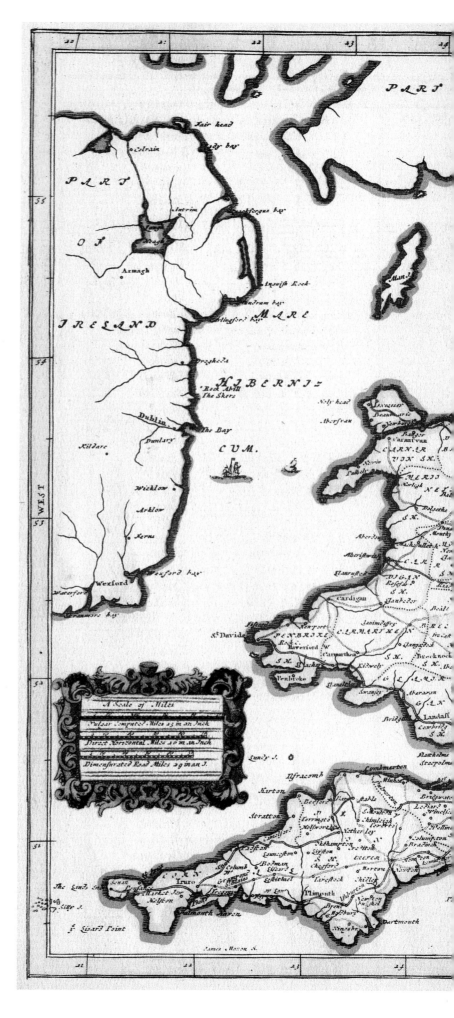

A New Map of the Kingdom of England from John Ogilby's *Britannia*, 1675, John Ogilby

Published in 1675, the *Britannia* atlas constitutes John Ogilby's most famous work, being one of the first examples of a comprehensive British road atlas. It largely comprises longitudinal strip maps of roads leading out of London and linking to other cities and towns. This map of the whole of England and Wales would be the first such survey since the Gough Map of circa 1300, and signaled the proliferation of such road maps in the years to come (John Adams would publish his renowned map *Anglia Totius* only two years later).

Ogilby had worked as a bookseller, a translator, a printer, a dancing master, geographer, and master of the King's Revels. He lost everything in the Civil War but, after the Restoration, managed to establish himself as a London printer. He was renowned in the field of cartography for revolutionising the printed road map. Ogilby was also responsible for standardising the mile to 1,760 yards at a time when many minor roads were still using a local mile.

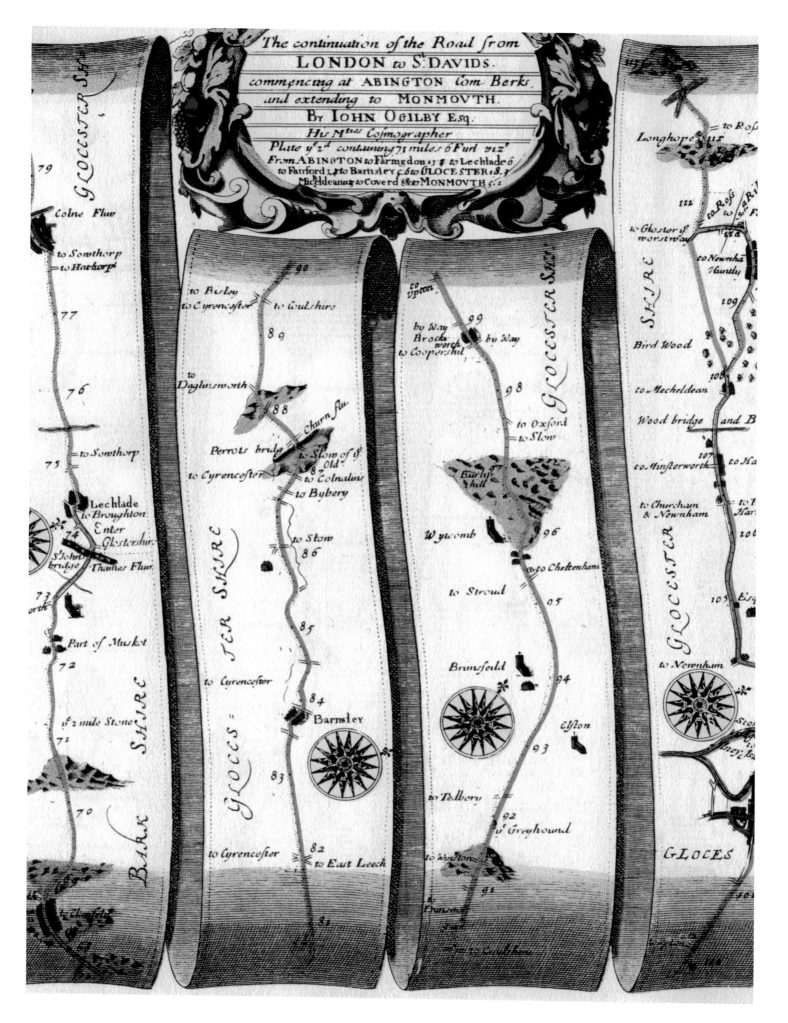

Ogilby had the foresight to make England seem reachable, even to those who had no intention of going anywhere. *Britannia* first appeared in the period between 1660 and 1707, when the country was officially "The Kingdom of England and Wales", having just emerged from a brief six-year period as "The Commonwealth of Great Britain and Ireland". In 1707, England was, in turn, to become a dominant part of the bigger and far more complex United Kingdom of Great Britain. This was an epoch when the nation was in flux and national identity a moving target. The ensuing preoccupation with maintaining England's identity as a whole—providing reassurance that the country was inter-connected and totalising, rather than a series of isolated and disconnected communities—when it could have easily have broken into regional blocks, pays great service to the power of maps such as Ogilby's.

POST HASTE

As the Ordnance Survey gradually spread its triangulation net across the country, the profits available to private mapmaking firms diminished, and their county mapping activities becoming incrementally superseded with a proportional decrease in demand for their maps. Naturally, private firms turned to other map-selling markets and none offered value more clearly than road maps. Those mapmakers integral to the practice already had the information they required at their fingertips and, with some judicious relabelling and added emphasis, it was transformed into the likes of Laurie & Whittle's A New Map of The Roads of England And Scotland, with the Distances in Measured Miles from Place to Place, 1800, Brookes' Travelling Companion through England & Wales, 1812, and Wallis' New Travelling Map of England and Wales with part of Scotland, 1815.

The increase of traffic—particularly the mail coaches that first made an appearance in 1784 after a successful trial run between Bristol and London—on England's roads would require better mapping for them to operate effectively at the end of the eighteenth century. (This requirement became especially pressing after Rowland Hill's introduction of the Uniform Penny Post in 1840, which calculated postage charges in relation distance.) In 1794, the mapmaker John Cary was appointed Surveyor of the Roads for the General Post Office (GPO) and his Cary's New Itinerary, a survey of all the major post roads was produced for the GPO in 1798.

The GPO also used commercially available maps including Bowles' Road Directory through England and Wales—published by Bowles & Carver in 1796 and advertised as "being a new and comprehensive display of the roads and distances from town to town and of each remarkable place from London"—and Moggs' General Map of the Roads of England and Wales, an improved edition of Paterson's Roads of 1829. In 1840, the GPO commissioned their own Circulation

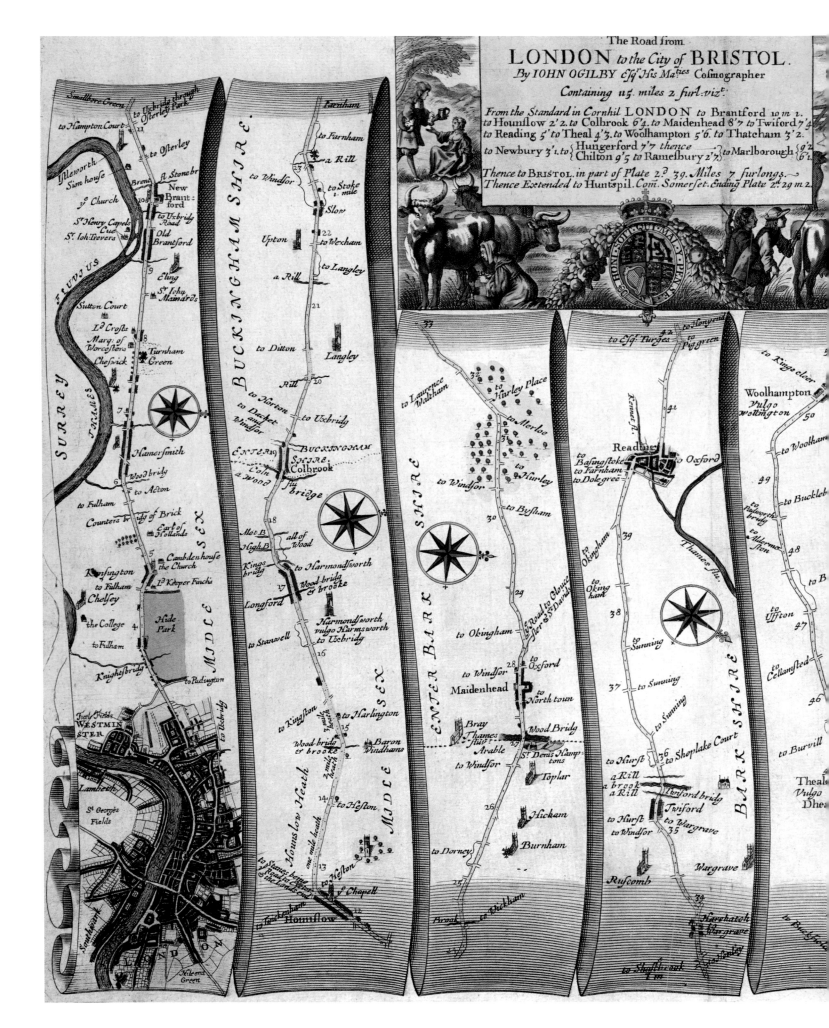

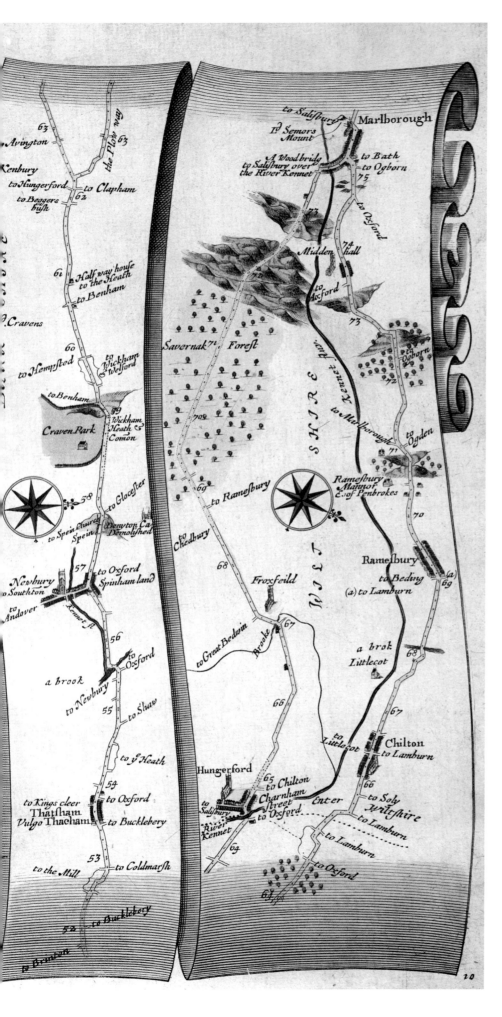

Road from London to the City of Bristol, from John Ogilby's *Britannia*, 1675, John Ogilby

Courtesy Bridgaman Art Library

A beautifully rendered, hand-coloured, engraving that reads from left to right, and bottom to top. The route itself leads out of the dense conurbation of London, along the Thames to Old Brentford, crossing it again at Maidenhead, then through Reading and into Berkshire, and then to Bristol on the subsequent page.

Great North Road out of London, 1790, John Cary

Courtesy Private Collection/
Bridgeman Art Library

This map is presented in two parts: the first depicts the road from London to Highgate and Hampstead and then on to Hendon; the second follows the journey on to St Albans. The cartographer, John Cary, was a prolific mapmaker and publisher in the eighteenth century. This map was part of his *Travelling Companion*, designed as a pocket atlas. The Great North Road now forms part of the A1, and was a similarly busy route during the eighteenth century, serving as a major coaching route. Many of the inns still survive along the road that linked London, York and Edinburgh. The map notes these inns and public houses to provide not only reference points for travellers but to offer places of rest and, possibly liquid recuperation.

maps of England and Wales, Ireland and Scotland from the mapmaker Edward Stanford, a firm that drew much of its business from supplying governmental and other official bodies.

MANIAS

Much of the enthusiasm displayed for transport in the eighteenth and nineteenth centuries would be characterised by the moniker 'mania'. In the two decades from 1750 to 1770 this 'mania' was unleashed at turnpikes (or toll roads), and over 400 acts of parliament were passed to set up independent turnpike trusts to develop and maintain the main roads system. By 1770, canals took over the mania mantle and, during the next 60 years, over 4,000 miles of canal were dug and put into use. 'Rail mania' exploded in the 1830s, reaching a peak in 1846 when 272 separate Acts of Parliament establishing railway lines were passed (although such frenzy almost completely disappeared by the end of the 1840s). Each mania required maps to assist the planning for the furious levels of destruction and creation involved: maps to run and, above all, to promote these new services, all with hope that they would appeal to the buying customer whose custom might help cover the extraordinary levels of cost involved in such investments.

Each of the systems of road, water and rail—joined later by air connections—was too complex to be effectively communicated on combined maps of the whole country so each ended up with their own. The Victorians, in particular, produced innumerable maps showing the rapidly emerging networks of lines spreading across the country. Connectivity was the primary emphasis of such maps, that wherever you are in the country you are just a short distance from the great arterial system. At any moment you could integrate with this great life-force traversing England; become an essential part of the nation or retreat to the suburban domesticity that you call home. It is these very opposing attractions that were dramatised by Noel Coward in *Brief Encounter*, and which still lie at the heart of the English love affair with the various transport systems installed over the years.[25]

However, following Henry Beck's development of the London Underground diagram from 1933 on, all of these systems have also developed their own, purely diagrammatic, representations; sometimes evoking the shape of the country, sometimes not. In their relative graphic simplicity the web-like skein of interconnectedness is frequently lost, and replaced by a metaphoric quality closer to that of plumbing or clinging ivy. No longer is the country bound tightly together by systems running through it; instead it appears to be the 'bones' of England itself, that is, supporting the infrastructure. Perhaps the notion of transport metaphorically, as well as literally, linking together the country has had its day. A new metaphor will be arriving shortly, we apologise for any delay caused.

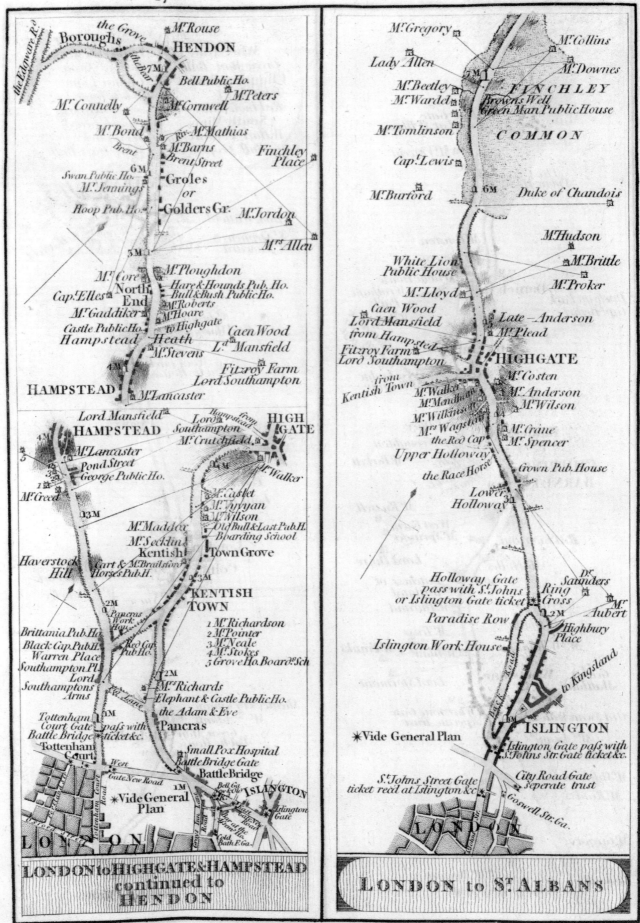

27

the Grove
the Edgware Rd
Boroughs
HENDON
Mr Rouse
the Bear
7 M
Bell Public Ho.
Mr Connelly
Mr Cornwell
Mr Peters
Mr Bond
Riv Mr Mathias
Brent
Mr Barns
Brent Street
Finchley
Place
Swan Public Ho.
6 M
Mr Jennings
Groles
or
Golders Gr.
Hoop Pub. Ho.
Mr Jordon
5 M
Mrs Allen
Mr Core
Mr Ploughdon
Cap.t Elles
North
End
Hare & Hounds Pub. Ho.
Bull & Bush Public Ho.
Mr Gaddiker
Mr Roberts
Mr Hoare
Castle Public Ho.
to Highgate
Caen Wood
Hampstead Heath
Mr Stevens
La Mansfield
Fitzroy Farm
Lord Southampton
HAMPSTEAD
4 M
Mr Lancaster

Lord Mansfield
Hampstead from Lord Southampton
HIGH GATE
4 M
Hampstead
Mr Crutchfield
Mr Lancaster
Pond Street
George Public Ho.
5
Mr Walker
3 2
Mr Castle
1
Mr Greed
Mr Vyvyan
Mr Wilson
13 M
Mr Maddox
Old Bull & Last Pub H.
Boarding School
Mr Seckline
Kentish
Town Grove
Haverstock Hill
Cart & Mr Brailsford
Horses Pub H.
3 M
KENTISH TOWN
2 M
Pancras Work Hou.
1 Mr Richardson
2 Mr Pointer
3 Mr Neale
4 Mr Stokes
5 Grove Ho. Boardg Sch
Brittania Pub Ho.
Black Cap. Pub Ho.
Red Cap Pub. Ho.
Warren Place
2 M
Southampton Pl.
Mrs Richards
Lord Southamptons Arms
Elephant & Castle Public Ho.
the Adam & Eve
1 M
Tottenham Court Gate
pass with ticket &c.
Pancras
Battle Bridge
Tottenham Court
Small Pox Hospital
Battle Bridge Gate
West
Gate New Road
Battle Bridge
1 M
Bell Ga.
ISLINGTON
Vide General Plan
Islington Gate
LONDON

Vide General Plan

28

Mr Gregory
Mr Collins
Lady Allen
Mr Downes
7 M
FINCHLEY
Mr Beetley
Mr Wardel
Browns Well
Green Man Public House
Mr Tomlinson
COMMON
Cap.t Lewis
Mr Burford
6 M
Duke of Chandois
Mr Hudson
White Lion Public House
Mr Brittle
Mr Proker
Mr Lloyd
Caen Wood
Lord Mansfield from Hampstead
Late — Anderson
Mr Plead
Fitzroy Farm
Lord Southampton
from Kentish Town
HIGHGATE
Mr Costen
Mr Anderson
Mr Walker
Mr Mendham
Mr Wilson
Mr Wilkinson
Mrs Wagstaff
Mr Crane
the Red Cap
Mr Spencer
Upper Holloway
the Race Horse
Crown Pub. House
Lower Holloway
Holloway Gate pass with St. Johns or Islington Gate ticket
Dr Saunders
Ring Cross
Mr Aubert
Paradise Row
2 M
Highbury Place
Islington Work House
to Kingsland
Vide General Plan
1 M
ISLINGTON
Islington Gate pass with St. Johns Str. Gate ticket &c.
St. Johns Street Gate ticket recd at Islington &c.
City Road Gate seperate trust
Goswell Str. Ga.
LONDON

LONDON to HIGHGATE & HAMPSTEAD continued to HENDON

LONDON to St. ALBANS

Published by J. Cary, July 1st 1790.

Bowles' Road Director through England and Wales: being a New and Comprehensive Display of the Roads and Distances from Town to Town and of each Remarkable Place from London, published by Bowles & Carver, 1796, Courtesy British Postal Museum and Archive/ Bridgeman Art Library

The importance of efficient and well-maintained post roads was celebrated in many maps of the late eighteenth century, not least this one by Bowles & Carver. Such renderings would dominate this field of mapping for many years until new systems of infrastructure were developed.

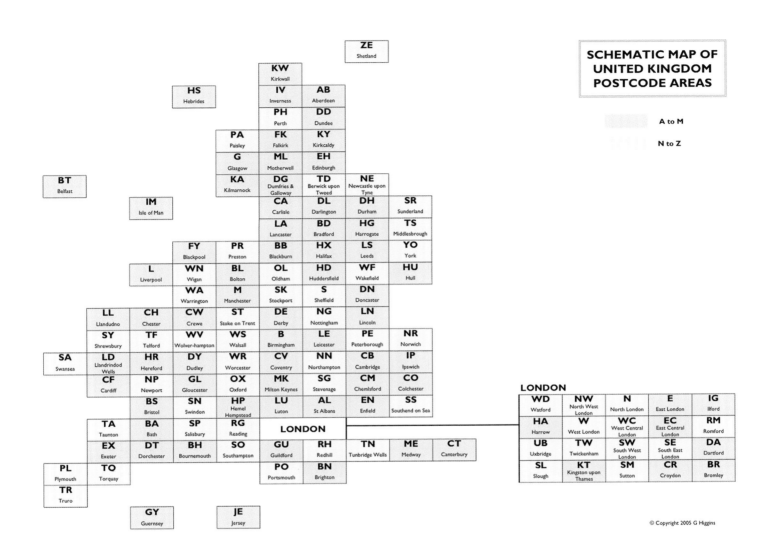

SCHEMATIC MAP OF UNITED KINGDOM POSTCODE AREAS

A to M

N to Z

© Copyright 2005 G Higgins

Schematic Map of United Kingdom Postcodes, Gerald Higgins, 2005

This schematic map of British postcodes, by Gerald Higgins, is an example of the manner in which freelance and creative individuals are taking mapmaking into their own hands—even with something so formal as the national address numbering system.

**Road Map, Finger Post Strip Map,
London to Exeter, circa 1900s,
Gerald Fothergill**

This is just one example of the many linear maps used by the Romans and later developed by Paris and Ogilby (amongst many others). Aimed at the new breed of road users—cyclists in particular—who were fanning out across the countryside during the century's new age of leisure, such maps are precise but provide no indication of the countryside surrounding the 'fingerposts' depicted.

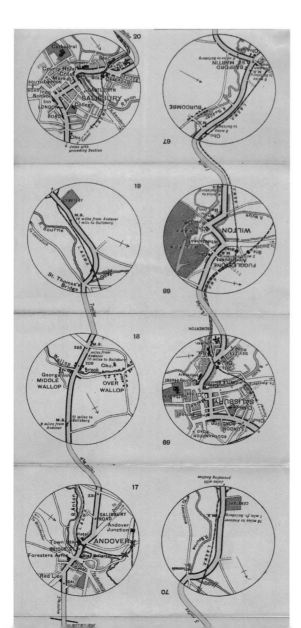

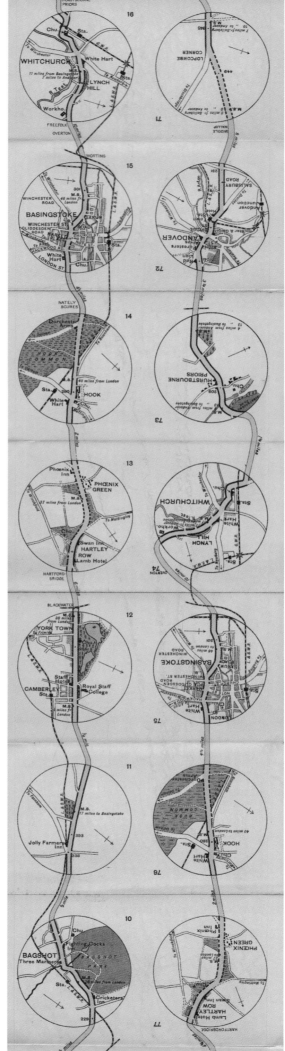

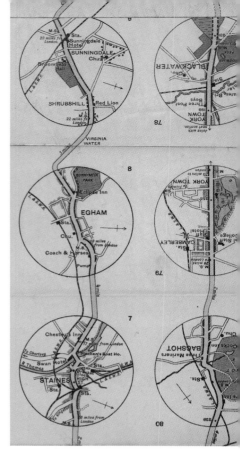

The map contains the following text elements:

The Distances of ye Principal Cities and Towns on the Crofs Roads from each other.

Meaf.d Miles		Comp.d Miles
74	Bristol to Banbury in Ox.	55
145	Bristol to West Chester	102
78	Bristol to Exeter	60
62	Bristol to Worcester	50
74	Bristol to Weymouth	57
80	Cambridge to Coventry	54
80	Carlisle to Barwick on Tweed	55
145	Chester to Cardiff in Glam	110
71	Dartmouth to Minehead in Som.	52
156	St Davids to Holywell in Flint	112
38	Exeter to Barnstable	32
79	Exeter to Truro in Cornwall	57
58	Glocester to Coventry	41
70	Glocester to Montgomery	50
86	Hereford to Leicester	67
71	Huntingdon to Ipswich in Suff.	54
43	Ipswich to Norwich	30
76	Kings Lynn to Harwich	59
42	Kings Lynn to Norwich	30
68	Monmouth to Lanbeder in Card	48
67	Notingham to Grimsby in Lincoln	50
68	Oxford to Bristol	48
80	Oxford to Cambridge	52
80	Oxford to Chichester	60
87	Oxford to Derby	68
87	Oxford to Pool in Dorset	61
69	Sinmouth to Carlisle	50
88	York to Lancaster	68
106	York to West Chester	74
43	Chelmsford to St Edmundsbury	35
52	Shrewsbury to Holywell	39
43	York to Scarborough	31

T. Badeslade delin.

W. H. Toms Sculpt.

Publish'd by the Proprietor W. H. Toms Sept. 29. 1742.

A Map of the principal Crofs-Roads from one Great Town, to another; thro: ENGLAND and WALES.

A Map of the Principle Cross Roads from One Great Town, to another, 1742, Thomas Badeslade

This printed atlas by Thomas Badeslade, and engraved by William Toms, curiously shows only the cross roads (rather than the principal roads evident in another map in the same atlas) connecting towns in England and Wales. It represents a significant decrease in the level of information from John Adams' map of 1679 and may never have been used for practical purposes.

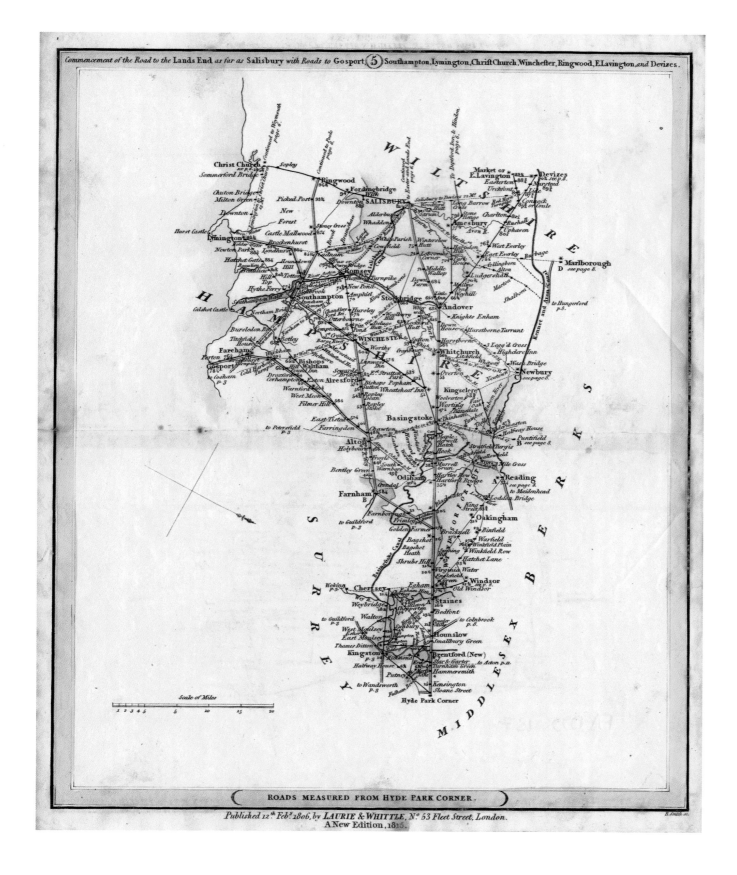

Commencement of the Road to the Lands End as far as Salisbury with Roads to Gosport ⑤ Southampton, Lymington, Christ Church, Winchester, Ringwood, E. Lavington, and Devizes.

ROADS MEASURED FROM HYDE PARK CORNER.

Published 12th Feb.y 1806, by LAURIE & WHITTLE, N.o 53 Fleet Street, London.
A New Edition, 1815.

Road Map, London to Gosport, Southampton, etc, Hampshire and Salisbury, etc, Wiltshire on the way to Lands End, from a Road Book by Laurie and Whittle 1806, edition of 1815

In this map from the *New Travellers Companion* by Laurie and Whittle, distances along roads are noted from Hyde Park Corner to the nearest quarter mile. The roads are graded in order of importance (the most important represented with a double line, and the lesser important with a doted line). The map is arranged in an upwards direction, therefore London is situated at the bottom and Portsmouth at the top.

It also features areas that straddle different counties in order to show the various coach and post routes between the town routes. All the roads are depicted in a linear manner, but overall accuracy in the distance marking and the positions of roadside inns has not been hindered as a result.

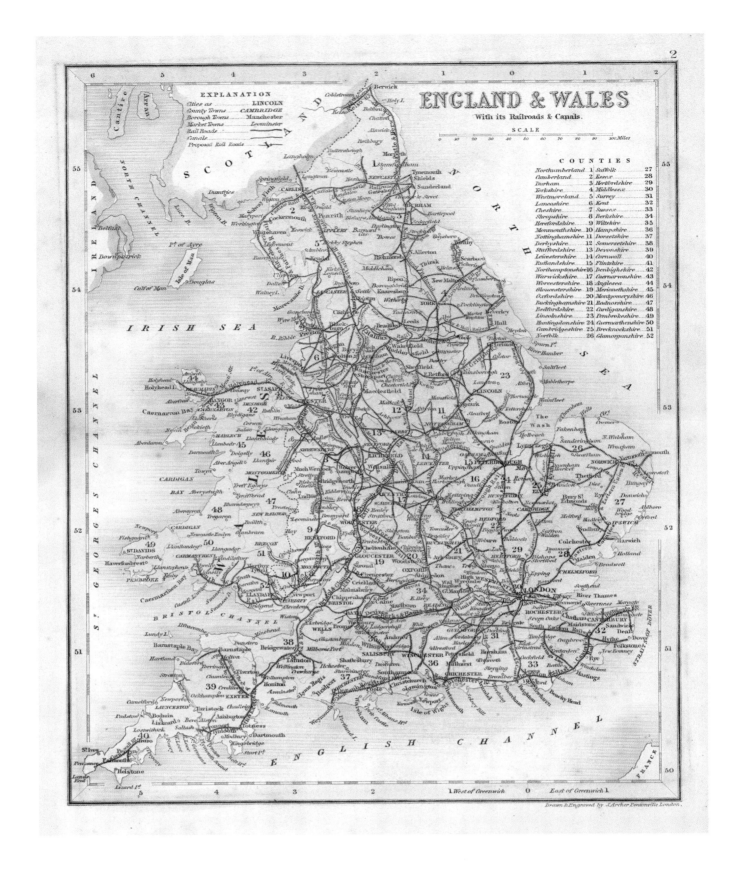

Map of the Canals and Railroads of England and Wales, circa 1857

Courtesy Mary Evans Picture Library

This map, published by the Society for the Diffusion of Useful Knowledge, depicts the network of canals and railroads that connect England and Wales. The Society was responsible for publishing inexpensive maps for general and widespread use in education, and also acted as the intermediary between authors and publishers on various publications. Their ethos was "the imparting [of] useful information to all classes of the community, particularly to such as are unable to avail themselves of experienced teachers, or may prefer learning themselves".

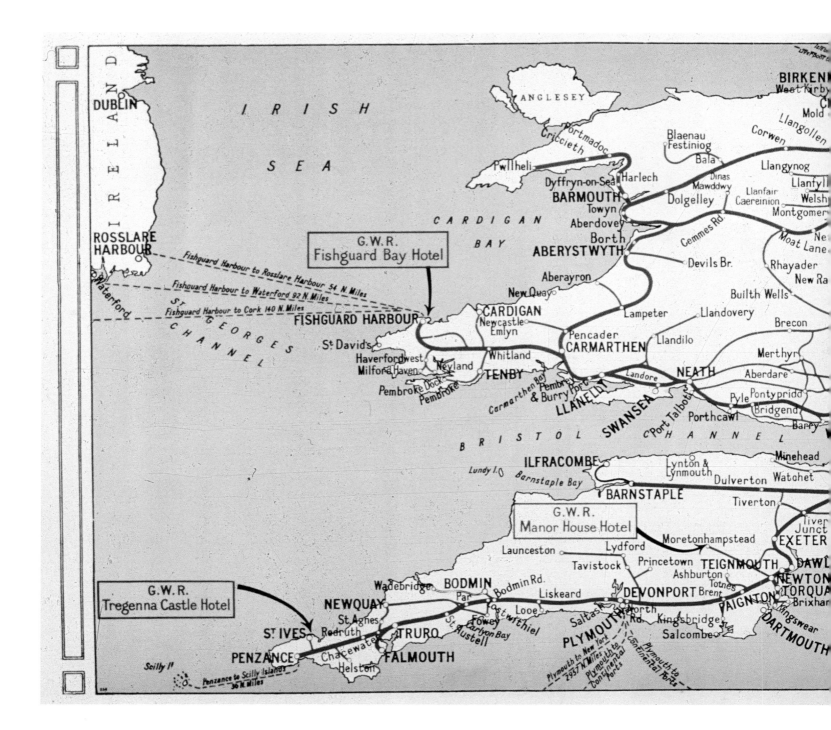

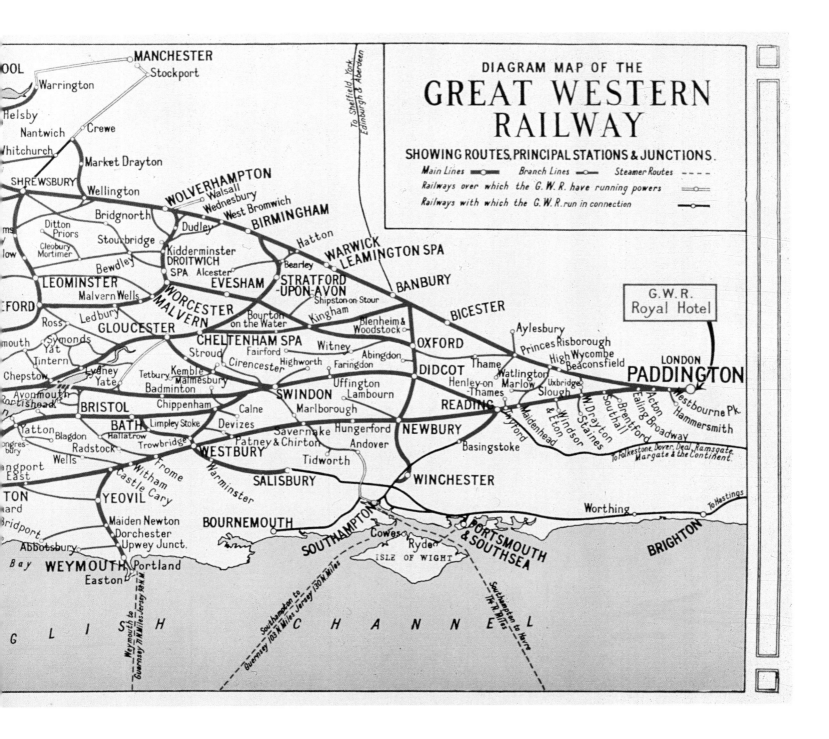

Great Western Railway Carriage Map, circa 1900

Courtesy Bridgeman Art Library

This map shows the principle stations and junctions for the passenger routes provided by the Great Western Railway. The Railway came into existence through an Act of Parliament in 1835 and was designed to provide a double tracked line from Bristol to London. The diagram shows Temple Mead and Paddington, the grand stations for Bristol and London respectively. At the turn of the century the Railway had expanded from a one-route system to a network of over 2,300 miles of track throughout England and Wales.

**London & North Western Railway
Poster of England and Wales,
early 1900s**
Courtesy NRM Pictorial Collection/
Science & Society Picture Library

This pictorial map of the railways sought to
encourage travel to several sites of natural beauty
and architectural importance. The map is carefully
arranged so that it appears to be emerging from the
smoke produced by the funnel of the steam train in
its bottom left-hand corner.

LONDON & NORTH WESTERN RAILWAY.

British Railways' Complete Passenger Network, 1961

Clement Attlee's Labour government sought to nationalise public services in the 1940s. In 1948 this led to the nationalisation of the 'Big Four' railway companies (Great Western Railway, London and North East Railway, Southern Railway and London, Midland and Scottish Railway) into 'British Railways'. This map shows the entirety of the British Railways network. The huge increase in the popularity of train travel is demonstrated by the multitude of routes depicted on the map, necessitating 2,000 new diesel locomotives to cover the rail system.

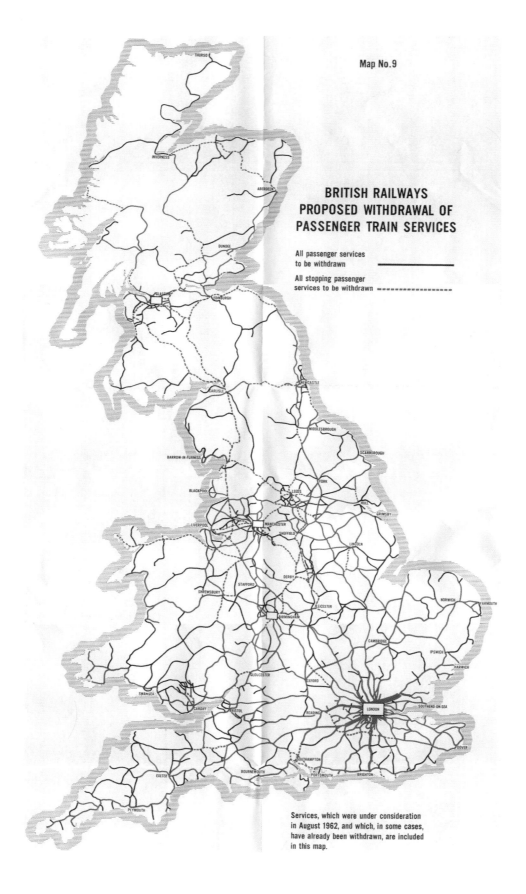

BRITISH RAILWAYS PROPOSED WITHDRAWAL OF PASSENGER TRAIN SERVICES

All passenger services to be withdrawn ——————

All stopping passenger services to be withdrawn ·····················

Services, which were under consideration in August 1962, and which, in some cases, have already been withdrawn, are included in this map.

British Railways Proposed Withdrawal of Passenger Train Services, 1963

This map presents British Railways' proposed withdrawal of passenger train services. Further modernisation of British Railways was intended in the 1950s, however, many rural railways were redundant, leading to their imminent closure.

This map was published in 1963, in conjunction with the Beeching Report, which resulted in the nationwide closure of passenger services over the next few years.

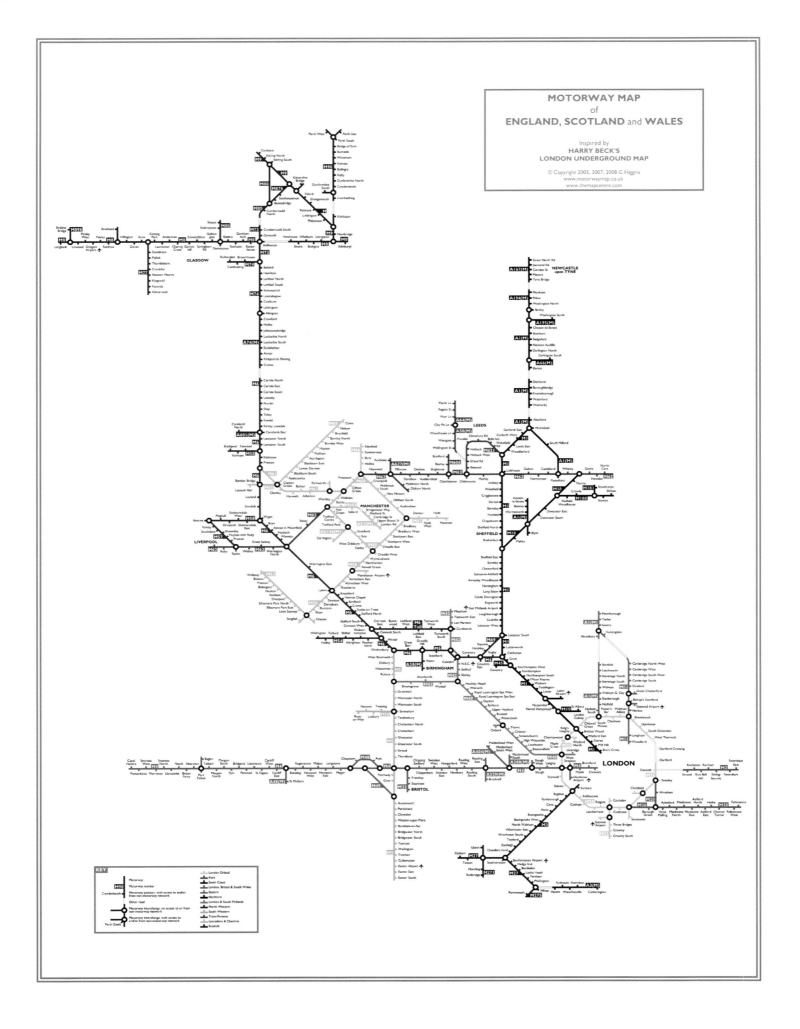

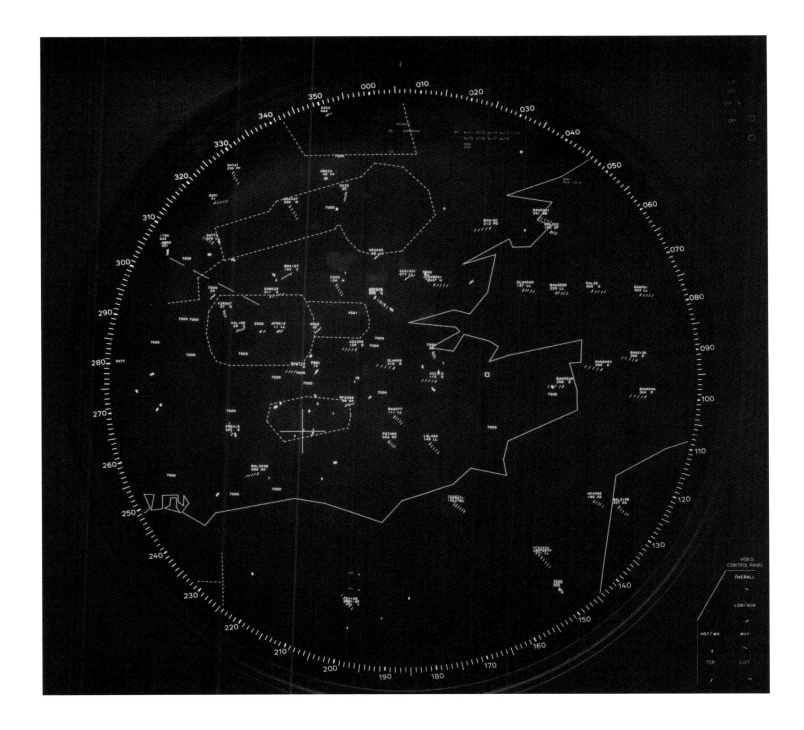

Opposite **Motorway Map of England, Scotland and Wales, Gerald Higgins, 2005**

The Moterway Map was inspired by Harry Beck's London Underground Diagram. Dissociated from the shape and distances of England or Britain, this map is free to define its own existence, generated by the geometric rules applied to its basic pattern.

Above **Air Traffic Control Display Screen at West Drayton, London, July 1992**

Courtesy Science Museum/
Science & Society Picture Library

This map of the air traffic control display at West Drayton depicts the mapping of air travel over England.

IMAGINING ENGLAND

THE IMAGE OF ENGLAND

PARADISE ON EARTH

> That spot to which I point is Paradise, Adam's abode, those lofty shades his bower.[1]

The conviction of the English that their country is blessed—a new Jerusalem and a sullied, but rebuildable, paradise—may have archaic roots, but its obvious harbinger was the English Reformation and the propagandist influence of Thomas Cromwell, Chief Minister to Henry VIII. The work Cromwell invested in re-inventing the English state, its church and history followed the country's renouncement of Catholicism (and, subsequently, its association with Rome) and resulted in the idea that England was a country set apart from the rest. This concept is one that would underpin the words penned by Blake in "Jerusalem", prose sung by Suffragettes and Rugby supporters alike:

> And did the Countenance Divine
> Shine forth upon our clouded hills?
> And was Jerusalem builded here
> Among these dark Satanic Mills?[2]

Cromwell set the context and commissioned and provided support for a large number of writers—Stephen Gardiner, Thomas Starkey,

Robert Fabyan, Richard Grafton and the mapmaker, John Speed, among them—to rewrite the story of England, in ways very familiar to us from a number of totalitarian twentieth century regimes. However, it was the arrival of the incendiary priest and writer, John Foxe (who had been exiled to Basel following the death of Mary I in 1559), that would revolutionise perceptions of the country's identity.[3] *Actes and Monuments* (or *Foxe's Book of Martyrs* as it is better known) was translated into English and went through five editions before the end of 1600. It was a virulent anti-Catholic history, encompassing a wide range of historical and not-so historical material.[4] As Edwin Jones states *Foxe's Book of Martyrs* was "one of the greatest and arguably the most successful of the world's works of propaganda" and a "reconstruction of the English past to suit the eyes of the Protestant present".[5] In *Britons: Forging the Nation 1707–1837,* Linda Colley points to the book's later influence and how it had a "new and much wider period of fame and was interpreted in a far more aggressively patriotic fashion in the eighteenth century and after".[6] *Foxe's Book of Martyrs* was the one book, along with the Bible, that could be found in the majority of homes in England for several centuries after its first publication. The potency of the myth of England it developed and inspired in its wake was profound. Originating in the aftermath of the English reformation, its spirit was re-invoked by Milton both in support of the republicanism of the sixteenth century Commonwealth of England and to bind the nation after the Glorious Revolution of 1688. If the spirit of *Foxe's Book of Martyrs* reaches its greatest poetic intensity in William Blake in whose "epic poems, England becomes the holy land, the seat of the ancient patriarchs and the home of the chosen race"; it achieves its apotheosis in the Victorian-Edwardian imperial heyday and the words of AC Benson in "Elgar's Pomp and Circumstance March No 1".[7]

> Land of Hope and Glory,
> Mother of the Free,
> How shall we extol thee,
> Who are born of thee?
> Wider still and wider
> Shall thy bounds be set;
> God, who made thee mighty,
> Make thee mightier yet.[8]

That it was England, rather than its inhabitants, which was the focus of the patriotic re-invention of the country, has inevitably meant that mapmakers were required to play a significant part in the description of the land being celebrated and defined. The mapmakers commissioned by William Cecil in the sixteenth century—Rudd, Saxton, Speed—not only produced highly practical work, but also provided the country with a recognisable shape that would allow it to be captured in the imagination.

UTOPIA ET ARCADIA

A map of the world that does not include utopia is not worth even glancing at, for it leaves out the one country at which humanity is always landing. And when humanity lands there, it looks out, and seeing a better country, sets sail. Progress is the realisation of utopias.[9]

Both utopia and its rural counterpart, arcadia, were re-invented in England in the Tudor period. Utopia was the fictional, ideal, Atlantic island championed by Sir Thomas More in 1516, and Arcadia—a Greek pastoral ideal—conjured up by Sir Philip Sydney in *The Countess of Pembroke's Arcadia* in 1580. The influence of both has inspired many mapmakers to re-interpret England in idealistic form. More's *Utopia* was published with a frontispiece depicting the island realm and the central city of Amaurote (although it only hints at some of the other 53 utopian cities). Abraham Ortelius also produced a vision of utopia in 1596, which had a visual affinity to the Isle of Wight and was dominated by rivers rather than cities. A third, more recent, version can be found in Satomi Matoba's contemporary rendering of the concept in *Utopia*, 1998. The most prominent aspect in all these examples is the compact island nature of utopia, a direct parallel to England itself, so often perceived as an island despite the presence of Scotland. This island-like theme is inherent

Utopia, 1596, Abraham Ortelius

A version of More's Utopia, not included in his
Theatrum, this image by Abraham Ortelius is only
known in one version. On the back, Ortelius notes
"This is that Utopia, bulwark of peace, centre of love
and justice, best harbour and good shore, praised by
other lands, honoured by you who knows why, this,
more than any other place, offers a happy life." It
looks suspiciously like the Isle of Wight.

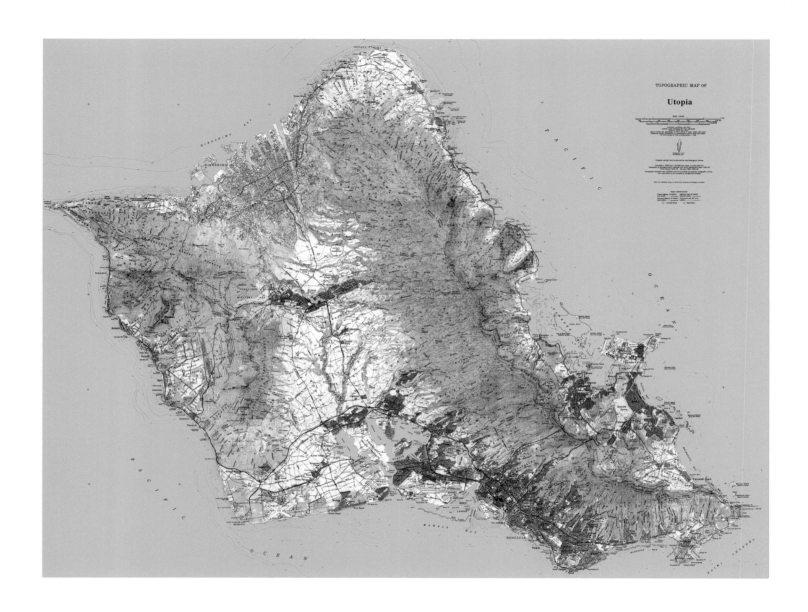

Utopia, computer manipulated map in an edition of 50, size variable, 1998, Satomi Matoba

Courtesy England & Co

Born in Hiroshima, Satoma Matoba moved to London to complete her studies in fine art. In her collage work, she combines familiar maps in order to present new and fictitious worlds. These in-between worlds signify her own position between Japan and London. *Utopia* assembles various landmasses to create a new 'perfect' landmass, furthering the project that More and his successors began. Through this and other works, Matoba explores what it means to be an immigrant, and interrogates whether or not immigrants will always be foreigners in their land of residence. In *Utopia*, she carves out the ideal territory, that third space, with which the diaspora can identify.

in maps of England across the centuries, and the country's fictional alter-
ego regularly makes an appearance in the literary imagination; Robinson
Crusoe's Treasure Island, Wildcat Island and San Seriffe. These maps
explore an ideal of England unhampered by political and social realities
but necessarily embody the spirit of their times.

The pursuit of arcadia has motivated mapmakers to similar
destinations as the inventors of imaginary islands. The arcadian, pastoral
ideal, combined with an English love of 'shaping' nature, led—via
the landscaped estates of Humphrey Repton and Lancelot 'Capability'
Brown—to the picturesque village that would be transplanted or
recreated in the wings of such landscapes. The idealised village, with
or without the worm in the bud that threatens to disrupt such fictional
locations as St Mary Mead or the Sussex village of Tilling, has its own
mapmaking tradition that is part of the cartography of England. That
so many of these maps live in the pages of fiction, especially children's
fiction, emphasises the role they play in shaping perceptions of England
as a whole—enticing a much wider audience than the traditional
navigational map.

Other maps are such shapes, with their islands and capes!
But we've got our brave Captain to thank:
(So the crew would protest) 'that he's bought us the best
—A perfect and absolute blank!'[10]

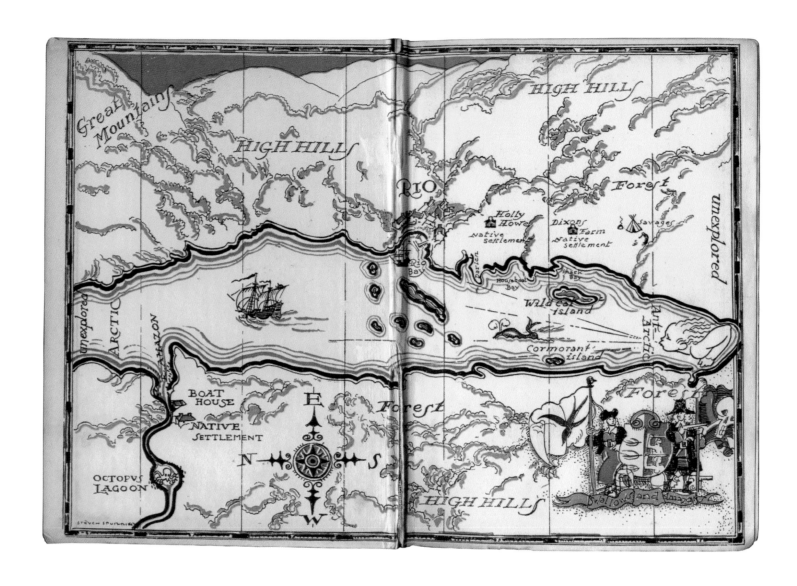

End-leaf map, Wildcat Island
from *Swallows and Amazons*,
1930, Arthur Ransome

Arthur Ransome's between-the-wars story of children setting up camp on a island in the Lake District is only one of innumerable tales of islands and escapism in English literature, whether in the south seas, in the tropics or in deepest England. They represent a search of an island race for a new utopia where they can start afresh and establish an English way of life unencumbered by history or the inconveniences of city life. Ransome's own illustration for *Wildcat Island* emphasises the romantic, as well as playing with the traditional idea of the pirate map, inevitably one of the themes of the book.

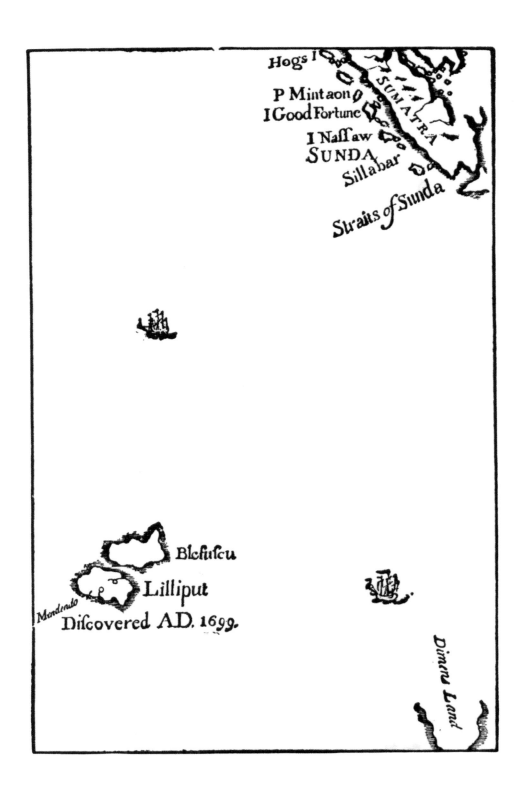

Hogs I

P Mintaon

I Good Fortune

I Naſſaw

SUNDA

Sillabar

SUNATRA

Straits of Sunda

Blefuſcu

Mandendo Lilliput

Diſcovered AD. 1699.

Dimens Land

Map, Lilliput and Blefuscu,
from *Gulliver's Travels*,
Jonathan Swift, 1726

Swift, picking up from Daniel Defoe's *Robinson Crusoe*—about a man shipwrecked on a small island—went several better by having the hero of his full-length satire, Samuel Gulliver, wash up on island after island, each stranger than the next. Lilliput and Blefescu are the islands of the tiny inhabitants, a twelfth of the size of normal humans, who first tie Gulliver down and later lionise him at the Lilliput court. The two islands are at war over which end a boiled egg should be cracked open. The islands become metaphors for the small-minded behaviours of their inhabitants, standing in for the English and their quarrelsome ways with the French and anyone else who challenged English authority.

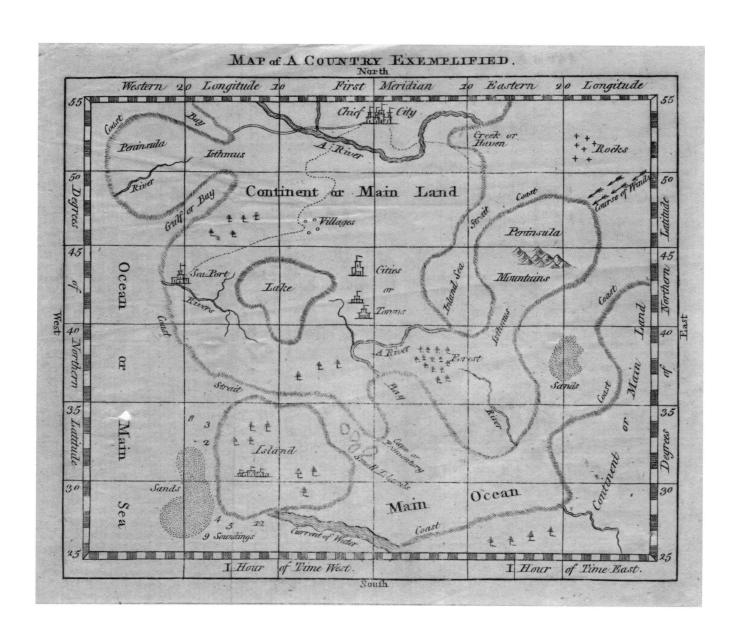

Map, uncoloured engraving, Map of a Country Exemplified, a fictional map for teaching, showing map features, circa 1700

The fascination with the map of imagined places that is so strong in the English mapping tradition is given innocent shape in this teaching aid, with its varied examples of different geographic features. It is a place ready for an adventure to begin, an 'X' to mark the spot and an expedition to be planned.

The Bellman's Map, from "The Hunting of the Snark" by Lewis Carroll, Henry Holiday, 1876

Possibly the culmination of the fantasy map, "a large map representing the sea / without the least vestige of land". In the poem Carroll touches on a cartographic unmentionable, that despite their ubiquity in Victorian times, and the national pride and expression of imperial identity that they embody—not everyone can read a map. The multifarious crew in search of the elusive Snark "were very much pleased when they found it to be a map they could all understand".

HERITAGE AT LARGE

If the desire to map a different, much-improved England led to the imagining of new islands and places, it also drove cartographers back to the past, resulting in maps that the Romans, Saxons or Medieval scribes might have produced if they had had the inclination and capacity. Such antiquarian visions of England are as potentially telling about the period in which they were made as fictional representations. The choice of presenting one past over another and the utililisation of history for the purposes of aggrandisement or propaganda was as current then as it is now. In particular, mapmakers would focus on the two periods of Roman rule and the Anglo-Saxon heptarchy. Both periods completely redefined the country's borders and capture a time when England was 'England' even if, inevitably, it was called something else.

The dominant figure in much of this work was the Rev Dr William Stukeley, 1687–1765, an antiquary, druid and priest, and one of the founders of field archeology in England. He undertook excavations of both Avebury Rings and Stonehenge and published extensive works on both (despite misdating them by several thousand years). Stukeley's reconstructions of both Roman and Saxon Britain were initially based on his observations on the ground but, in 1747, he received a letter from a student, Charles Bertram, describing the discovery of a Medieval copy of a Roman map, the *De Situ Brittaniae*, in Denmark.

Stukeley used a facsimile of this map as the basis of a study (in which he identified the author as a fourteenth century monk—Richard of Cirencester) that would inform much of his subsequent work. The map was a hoax, and borrowed material from a number of sixteenth century sources, a fact that would remain undiscovered until 1845, when German historian, Karl Wex, published a definitive refutation. By that time, Bertram's relatively crude forgery had found its way (via Stukeley's) 'scholarship' into the standard maps of both Roman and Saxon England and onto many Ordnance Survey maps. The Pennines are an invention of Bertram's, who labelled them as the "Pennines Montes", as were several other place names. The process of unravelling truth from fiction kept scholars and mapmakers engaged for long after Wex's debunking of Stukeley's unknowingly innocent efforts.

Despite this, the more rigourous archeological and historical maps that were produced after his foray into mapping, would still rely on much of the pioneering work he undertook. Mapping the land, especially when combined with specialist satellite and aerial photography, has revealed information about the past that would not have been possible any other way. Mapping and archeology have become deeply interdependent disciplines after a difficult and contentious start in life together.

Road Map, the Antonine Itineraries through Britain, 1723, William Stukely

The Reverend William Stukeley, 1687–1765, was an antiquary and pioneer of archaeology in England, carrying out the first investigations at Stonehenge and Avebury. In particular, Stukeley combined his interest in freemasonry, Druidical religion and his own English protestantism in a heady brew that is behind many of the myths that still cling to the early history of the British Isles. Despite his status as an ordained priest, he was familiarly known as the 'Arch-Druid'.

In his bid to create a 'universal history', Stukely produced many maps of early England, including this one of the Roman Antonine Itinerary, showing the stations and settlements of the Roman Empire in England and Wales. Stukeley developed the extensive use of maps to show factual information about the past, as much as others were doing with the present, but he also used them for promoting a mythologised version of the past that has now become deep-seated in the English self-image.

Britain as it was Devided in the Tyme of the Englishe Saxons, 1616, John Speed

Courtesy Stapleton Collection/
Bridgeman Art Library

Created in 1616 by John Speed, this hand-coloured map of the British Isles shows the Anglo-Saxon heptarchy with the 14 vignettes presenting fanciful images of the Saxon and Anglo-Saxon Kings from Hengist on. The idea of the heptarchy was developed in the twelfth century by Henry of Huntingdon to cover the years 350 to 850 AD, when history was hazy and uncertain. The heptarchy became common in the sixteenth century, following the nation-building efforts of Thomas Cromwell for Henry VIII. Speed's map is part of this consolidation of the past into an acceptable version for consumption by the public.

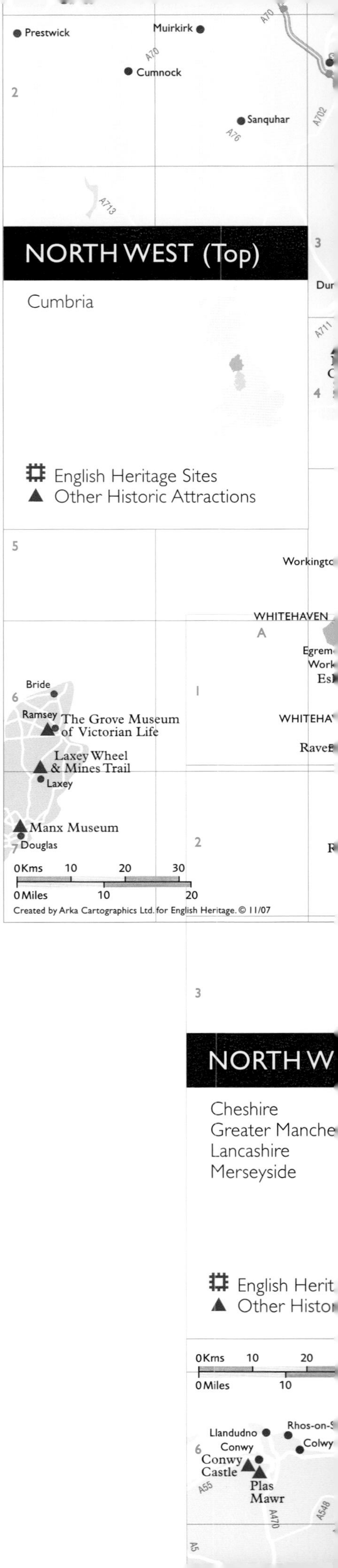

NORTH WEST (Top)

3

Dur

Cumbria

4

⊞ English Heritage Sites
▲ Other Historic Attractions

5

Workingto

WHITEHAVEN

A

Egrem
Work
Es

Bride ●

1

Ramsey ● The Grove Museum
▲ of Victorian Life

WHITEHA

Rave

Laxey Wheel
▲ & Mines Trail
● Laxey

6

▲ Manx Museum
7 ● Douglas

2

R

0 Kms   10   20   30
0 Miles   10   20

Created by Arka Cartographics Ltd. for English Heritage. © 11/07

3

NORTH W

Cheshire
Greater Manche
Lancashire
Merseyside

⊞ English Herit
▲ Other Histor

0 Kms   10   20
0 Miles   10

Rhos-on-S

Llandudno ●
6 Conwy   ● Colwy
Conwy ▲▲
Castle
Plas
Mawr

Right and overleaf

English Heritage map of England, circa 2008,

Arka Cartographics Ltd

Courtesy English Heritage

This English Heritage map is part of a 2008 map of England showing archaeological sites across the country—the past has never been more popular than it is at present.

The goal of the map is to enrich public understanding of how humans have inhabited the land as well as track the movement of past populations and settlements. Such information is invaluable, in that it reveals more and more about the characteristics of various historic cultures across the country. As such, greatest priority has been given to the areas that are under threat of erosion, building, or other activities that might disrupt the archaeological qualities of the region.

English Heritage Sites
Other Historic Attractions

East Riding of Yorkshire
North East Lincolnshire
North Lincolnshire
North Yorkshire
South Yorkshire
West Yorkshire

English Heritage Sites
Other Historic Attractions

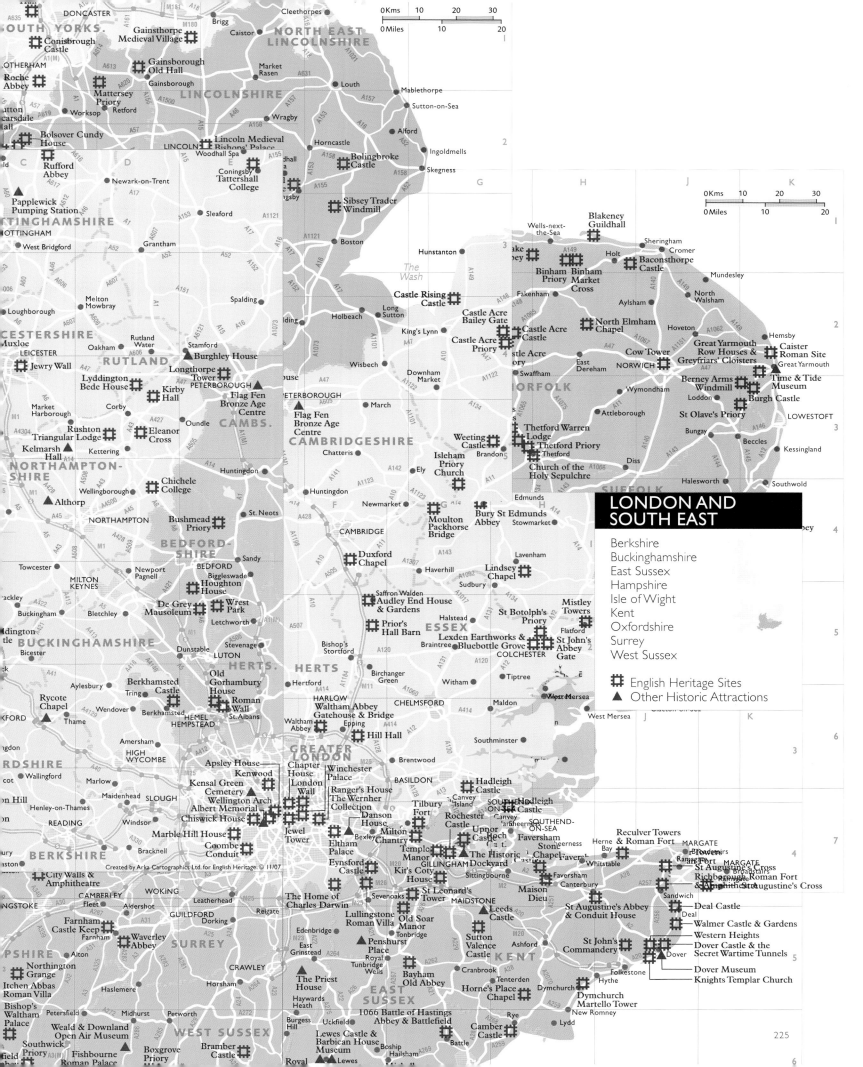

MAPPING ENGLAND

Jigsaw Map of Europe, 1766, John Spilsbury

Courtesy the British Library

John Spilsbury was the apprentice to the Royal
Geographer and the first commercial manufacturer
of jigsaw puzzles. Jigsaw puzzles depicting maps
were intended for use in schools so pupils could
learn—not only about specific geographies—but
also how countries interrelate. These maps were
not in the shapes familiar to us today; instead, the
pieces were based on the borders of countries,
oceans, lines of latitude and longitude, and other
similar intuitive limits.

PLAYING AROUND

The Victorians, ever keen to deploy 'playing' for didactic and pedagogical purposes, used children's games to teach geography as much they did to lead soldiers to inculcate military manoeuvres into young minds. Maps were ideal candidates for board games or jigsaw puzzles and games such as Wallis's Picturesque Round Game of the Produce and Manufacture of the Counties of England and Wales were designed to make the acquisition of knowledge (in this case the produce associated with 151 different towns and cities *en route* to the final destination of London) as painless as possible.

Playing cards had been produced for similar purposes from the seventeenth century on—William Bowes' set of 1590 and Robert Morden's from 1676 are two such examples—the 52 counties of England and Wales making an extraordinarily convenient pack.

Jigsaws of both England and the British Isles were initially hand-cut and made from wood, however, with the advent of better cutting technology and the development of cardboard as a cheap but effective backing material in the twentieth century, production really took off. Several generations probably learnt the basic geography of their country by such means. By the twenty-first century printing technology that allows almost any image to be converted into a jigsaw has seen a flurry of map-based possibilities (a particular focus currently being reprints of local Ordnance Survey maps).

In all their seriousness and educational intent, these games were not intended to entertain. It was only with the advent of role-playing and computer games at the end of the twentieth century that such games were perceived as 'fun'. Both these genres have had no problem with playing fast and loose, borrowing from parts of England and hacking and chopping away at it as they saw fit. The result has been the invention of many new 'Englands', 'Albions' and 'Britains' both on boards and in cyberspace. Some represent a foundational base for civilisation-building, while others present landscapes ripe for warfare and for other role-playing adventures set in any number of mythological and historical periods.

The most intriguing are those maps not necessarily intended to be seen by the ordinary player, those that represent the underlying structure of the game playing world and its different levels. The map of England becomes, in these images, most detached from reality. But it also expresses the multilayered quality of any good map—one that encompasses disparate strata of knowledge that is overlaid and interconnected. In its portrayal of the infinitely complex physical world, cyberspace embodies a new challenge to the practice of mapping and how maps are perceived. Developments in gaming will also allow significant breakthroughs in the complex nature of mapping the world— once it moves on from its current preoccupation with car chases.

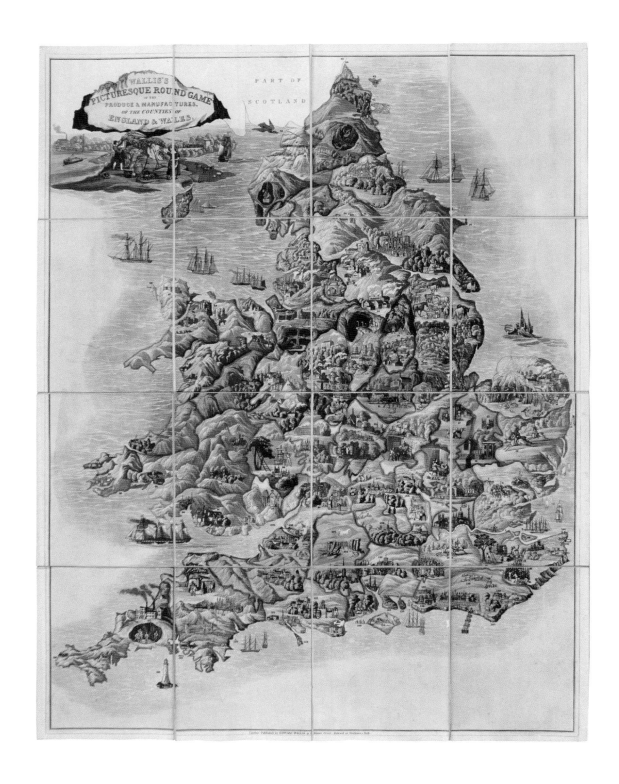

Wallis Picturesque British Map Game, circa 1850

Courtesy George Glazer Gallery

This map was designed to serve both educational and entertainment purposes. The game starts at the mouth of the River Thames and players travel around 151 different counties until finishing at London. It was intended to arm children with useful information on the features of each county as they played. These features include landmarks, topographical points of interest and areas of cultivation (for example, those well-known for the production of strawberries, such as Ely). There is a seriousness about such games that cannot have been very appealing to children, but they served a very useful purpose in at least familiarising a young age group with the shape and image of the country.

**The Cottage of Content or Right Road
and Wrong Ways Board Game, 1848,
William Spooner**
Courtesy V&A Images
A map leads the players through the warren of
temptation to temperance and ultimately the cottage
of contentment in this proselytising game for the
virtues of abstinence. It plays with the perception
of a traditional map and may have succeeded in
interesting more children in the idea of cartography
than that of teetotalism.

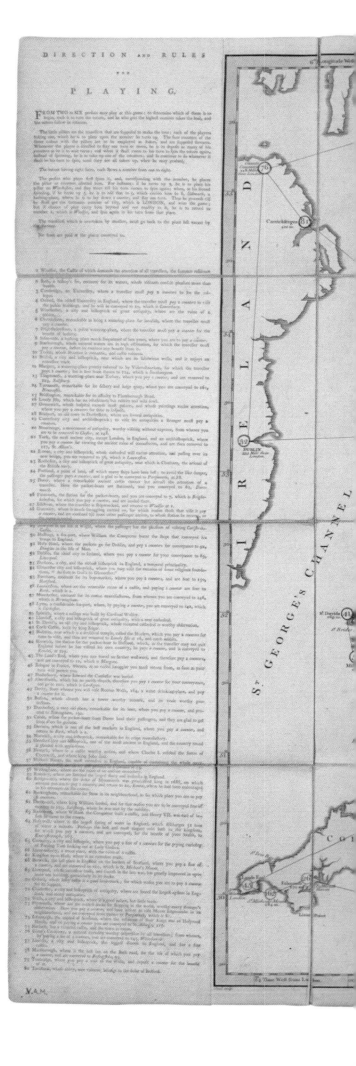

**A New Royal Geographical Pastime
for England and Wales Board Game, 1787,
Robert Sayer**

Courtesy V&A Images

The sub-title of this game is "Whereby the distance
of each town is laid down from London in measured
miles being a very amusing game to play with a
teetotum, ivory pillars and counters". It has the full
look and feel of an accurate map and was likely to
have been an enterprising effort by the mapmaker
to extend his market, while unusually providing
players with information that might actually prove
useful in the wider world. 169 principal and county
towns are joined by lines. The rules on either side of
the map list each town, and any rewards or forfeits
to be made when landing there. Stonehenge is
deemed "worthy of visiting without expense, from
whence you are to be removed to Chester at 148",
while Knaresborough "has four medicinal springs of
different qualities; to drink from them you must pay
one counter and are conveyed to Bath, 2".

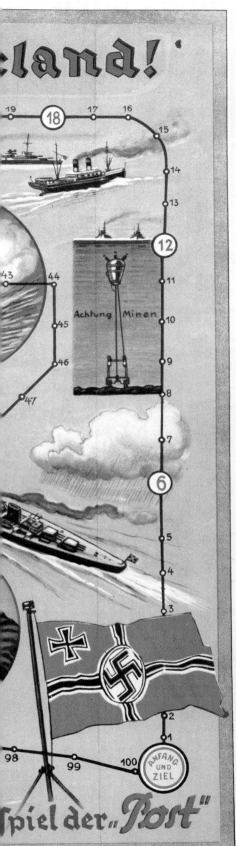

Spielregeln

An diesem interessanten Spiel können bis zu fünf Personen teilnehmen. Jeder Spieler erhält eine Figur, ein Kriegsschiff oder ein Kampfflugzeug zum Setzen auf den vorgezeichneten Gefechtsweg. Es wird mit einem Würfel gespielt. Wer zuerst sechs würfelt, beginnt. Er hat das Spiel zu leiten und die Spielregeln sowie die Erläuterungen vorzulesen. Die Reihenfolge der übrigen Spieler richtet sich nach der erzielten höchsten Augenzahl. Bei allen roten Nummern sind die Erläuterungen besonders zu beachten. Wer zuerst am Ziel eintrifft, das durch direkten Wurf erreicht werden muß, ist der Sieger und erhält den ausgesetzten Preis.

Erläuterungen zum Gefechtsweg

Nr.
6 Tiefe Wolkendecke, Schlechtwetterzone, alle Flieger müssen umkehren und nochmals vom Ausgangshafen starten. Die Schiffe verlangsamen die Fahrt und fahren auf Nr. 5 zurück.

12 Vorsicht! Minenalarm für alle Schiffe. Diese setzen einmal mit Würfeln aus. Die Flieger überfliegen dieses Gebiet und rücken auf Nr. 13 vor.

18 Deutscher Kreuzer bringt feindliches Handelsschiff auf, darf nochmals würfeln. Die Flieger beobachten und schützen den Kreuzer und warten auf Nr. 17 bis zum nächsten Wurf, das heißt sie müssen von Nr. 18 auf Nr. 17 zurück. U-Boot fährt auf Nr. 21 vor.

23 Deutsches Kampfflugzeug schießt englischen Jäger ab, darf auf Nr. 27 vorrücken. Die auf Nr. 23 ankommenden deutschen Schiffe und das Aufklärungsflugboot beteiligen sich an der Rettungsaktion der feindlichen Fliegerbesatzung und rücken auf Nr. 24 vor.

26 Deutsches U-Boot muß feindlichen Wasserbomben ausweichen, einen Umweg machen und auf Nr. 22 zurück. Die Flieger vertreiben den englischen Zerstörer, der die Wasserbomben geworfen hatte, und dürfen auf Nr. 27 vorrücken, ebenso der Kreuzer, der sich am Gefecht beteiligt hat.

30 U-Boot schießt Torpedo und versenkt feindliches Schlachtschiff, darf beim nächsten Wurf die doppelte Würfelzahl rechnen. Die Flieger machen Fotoaufnahmen und setzen einmal mit Würfeln aus.

33 Deutsche Flieger greifen im Hafen von Firth of Forth englische Schiffe mit Bomben an, dürfen auf Nr. 35 vorrücken. Die Schiffe erhalten Warnsignal, dürfen nicht auf Nr. 33 stehen bleiben, sondern müssen auf Nr. 32 zurück und Gefahr abwarten.

36 Deutsches Aufklärungsflugboot gerät in englisches Abwehrfeuer, muß auf Nr. 29 zurück. Kampfflugzeuge fliegen sehr hoch, kommen durch, auf Nr. 38. Auch die Schiffe machen schnelle Fahrt und kommen auf Nr. 37 vor.

39 Deutsches Aufklärungsflugboot sichtet feindlichen Flugzeugträger, meldet seine Wahrnehmung an die deutschen Schiffe, darf nochmals würfeln.

42 Deutsche Kampfflieger bombardieren englischen Flugzeugträger, dürfen auf Nr. 44 vorrücken. Die übrigen Flieger stoßen auf keinen Feind und setzen einmal mit Würfeln aus. Die Schiffe dürfen sich auf Nr. 42 nicht aufhalten, sie müssen auf Nr. 41 zurück.

49 Deutscher Kreuzer kommt in der Nordsee in ein Gefecht mit englischen Zerstörern, verliert dadurch Zeit, muß einmal mit Würfeln aussetzen. Auch die Flieger beteiligen sich am Kampf und müssen auf Nr. 48 zurück.

53 Achtung! Minensperre! U-Boot unterfährt die Sperre, darf auf Nr. 55 vor, der Kreuzer muß auf Nr. 51 zurück. Die Flieger überfliegen dieses Gebiet und rücken auf Nr. 56 vor.

58 Die deutschen Aufklärer über Liverpool, starkes Abwehrfeuer, setzen einmal mit Würfeln aus. Die Schiffe, welche die Nummern 58, 60 und 61 durch Würfeln erreichen, umfahren Schottland und kommen durch den Nordkanal auf Nr. 62, sie müssen aber dort zweimal mit Würfeln aussetzen.

61 Sämtliche Flugzeuge geraten bei einem Vorstoß auf das Industriegebiet bei Glasgow in starke Abwehr, sie müssen auf Nr. 60 zurück.

66 Deutsches U-Boot macht einen kühnen Vorstoß in den Bristol-Kanal, erledigt feindliches Torpedoboot, darf auf Nr. 70 vorrücken. Der Kreuzer darf auf Nr. 68 vorrücken. Englische Flieger starten in großer Übermacht vom Flughafen Bristol aus und drängen die deutschen Flieger auf Nr. 64 ab.

69 U-Boot wird vom Feinde überrascht, muß sich schleunigst auf Grund legen und Gefahr abwarten, setzt einmal mit Würfeln aus, der Kreuzer dagegen fährt weiter auf Nr. 70. Die Flugzeuge streifen über dem Kanal und gehen vor auf Nr. 72.

73 Deutsche Kampf- und Sturzkampfflugzeuge greifen die Festung Plymouth mit Erfolg an, sie dürfen nochmals würfeln.

76 Angriff auf Portsmouth geplant, alle Flugzeuge geraten im Kanal in Nebel, müssen abbiegen und auf Nr. 74 zurück. Das U-Boot rückt auf Nr. 78 vor. Der Kreuzer jedoch muß einmal mit Würfeln aussetzen.

83 Deutsches Aufklärungsflugboot erreicht London, umfliegt Ballonsperre und macht wichtige Fotoaufnahmen, darf nochmals würfeln. Die Schiffe können die Nr. 83 und 84 natürlich nicht erreichen, sie müssen auf Nr. 82 zurück.

85 Deutscher Kreuzer beschießt die englische Ostküste, darf auf Nr. 90 vorrücken. Sturzkampfflugzeug hat Motorstörung, setzt einmal mit Würfeln aus.

94 U-Boot wird bei Überwasserfahrt von feindlichem Flieger angegriffen, muß sofort tauchen und seine Fahrt verringern, setzt einmal mit Würfeln aus. Deutsche Kampfflugzeuge vertreiben das englische Kampfflugzeug und dürfen auf Nr. 96 vorrücken.

97 Sämtliche Schiffe und Flugzeuge geraten in Sturm, werden aufgehalten und müssen auf Nr. 95 zurück.

EnV̄ ta 26852

Figuren zum Ausschneiden

AUFKLÄRUNGS-FLUGBOOT

AUFKLÄRUNGS-FLUGBOOT

U-BOOT

U-BOOT

STURZ-KAMPFFLUGZEUG

STURZ-KAMPFFLUGZEUG

KREUZER

KREUZER

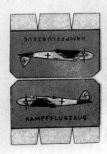

KAMPFFLUGZEUG

KAMPFFLUGZEUG

We're going to England! Board Game for Children based on the Naval and Air Battles for England, 1940, V Romer

Courtesy Musée National de l'Education, France/ Bridgeman Art Library

A German wartime children's board game celebrating the attack on England by the Luftwaffe and the Kriegsmarine. There is clear recognition of the neutrality of Eire even if the England/Britain issue remains unresolved. Only London is represented by the usual tourist imagery, elsewhere the pictograms are all strategic targets; docks, power stations, factories and transport infrastructure. It all seems a very jolly game of destruction.

Kingmaker, 1974

Produced in 1974, this board game is based on the fifteenth century Wars of the Roses between supporters of the Houses of Lancaster and York. It is largely a strategic game played on a map of England showing routes between the various towns, marked by important buildings or coats of arms with place names. Such games show the appropriation of maps by many other interest areas—very different from the Victorian pedagogical view—and the freedom with which they could be used.

Screenshot, Sid Meier's Civilization III, Firaxis Games, 2003

The computer game, Civilization (in its series of evolutions) allows players to start afresh with building complex societies in semi-familiar parts of the world. This screenshot from an ongoing game reveals an advanced and highly militarised 'Greek monarchy' establishing itself in 1380 AD, in what appears to be Southern England, although Hastings, Canterbury and Gloucester are not entirely in their geographically familiar locations. Computer games, often very reliant on underlying 'maps' of cyberspatial territory, have done more to make maps common property and re-invigorate the potential for cartographic innovation than any other medium.

THE SELF-CONSCIOUS IMAGE

Over the centuries, mapping has been used to cultivate an image of England, by those seeking both to praise and damn the country as well as those simply wishing to amuse a fickle audience. One of the earliest examples includes a copy of the Roman Imperial Chancery document *Notitia Dignitatum, Dux Britanniarum*, which shows fourth century Roman forts and garrisons spread from England across to Hadrian's Wall. This non-graphical depiction sought to communicate military might and control. Heinrich Bunting's sixteenth century 'map' indicates the opposite. A cloverleaf in which Jerusalem plays the central role makes clear that 'Engeland' has barely a role of which to speak, as it sits on the far edge of Europe, with little hint of any civilisation.

It is only when cartoonists appropriate the map of England for their own purposes, when the idea of combining country, map and person together to create a scabrous image of a nation behaving badly, that the idea of manipulating such imagery really takes off. But, despite a basic distortion and transformation of the shape of the country, these images do not substantially change the physical make up of England. Outlines are used as shorthand for either country or state, but the key elements are still retained.

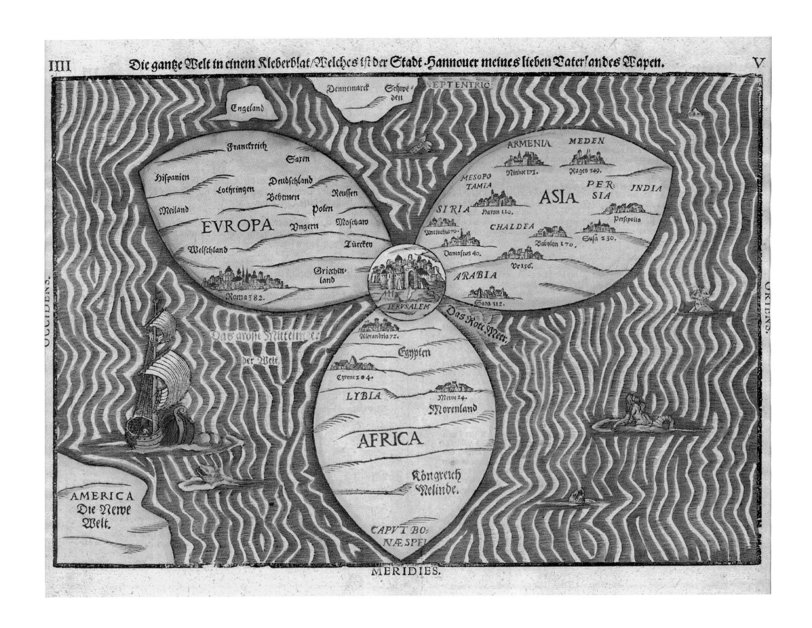

Die gantze Welt in einem Kleberblat/Welches ist der Stadt Hannouer meines lieben Vaterlandes Wapen.

IIII

V

SEPTENTRIO.

Dennemarck   Schwe-
                den

Engeland

Franckreich   Saxen   ARMENIA   MEDEN

Hispanien   Deudschland   Ninive 171.   Nages 349.
          Lothringen   Reussen   MESOPO
  Meiland   Behmen   Polen   TAMIA   ASIA   PER-   INDIA
                          Moschaw   SIA
          EVROPA   Vngern   SIRIA   Haran 110.   Persepolis
    Welschland   Türcken   Antiochia 70.   CHALDEA   Susa 230.
                  Griechen-   Damascus 40. Babylon 170,
                    land   Ur 156.
        Roma 382.   ARABIA
                                      JERVSALEM   Saba 112.

Das grosse Mittelmeer.   Das Rote Meer.
              der Welt.   Alexandria 72.

                              Egypten

                  Cyrene 284.
                LYBIA   Meroe 24.
                              Morenland

                        AFRICA

AMERICA   Königreich
Die Newe   Melinde.
  Welt.

                              CAPVT BO-
                              NÆ SPEI

                    MERIDIES.

OCCIDENS.   ORIENS.

The Whole World in a Cloverleaf, 1581, Heinrich Bunting

Heinrich Bunting enjoyed creating maps that took on symbolic shapes and he was certainly more worldly than to believe in the old T-O map view of the world that has Jerusalem at the centre. However, this map shows his playful re-invention of that traditional form, based on a seventh century example created by Saint Isidore, using the shape of a cloverleaf, the heraldic symbol of his home town, Hanover. Europe, Asia and Africa spin off from the centre and he has included random labels of countries, cities and people across each of the continents as appropriate.

But England has a special place in the mapmakers imagination, clearly not part of the cloverleaf, and even coloured separately. As exotic as one of the fish or mermen who swim in the sea around the leafshape—England is separate and easily identifiable, yet somehow awkward and difficult to place.

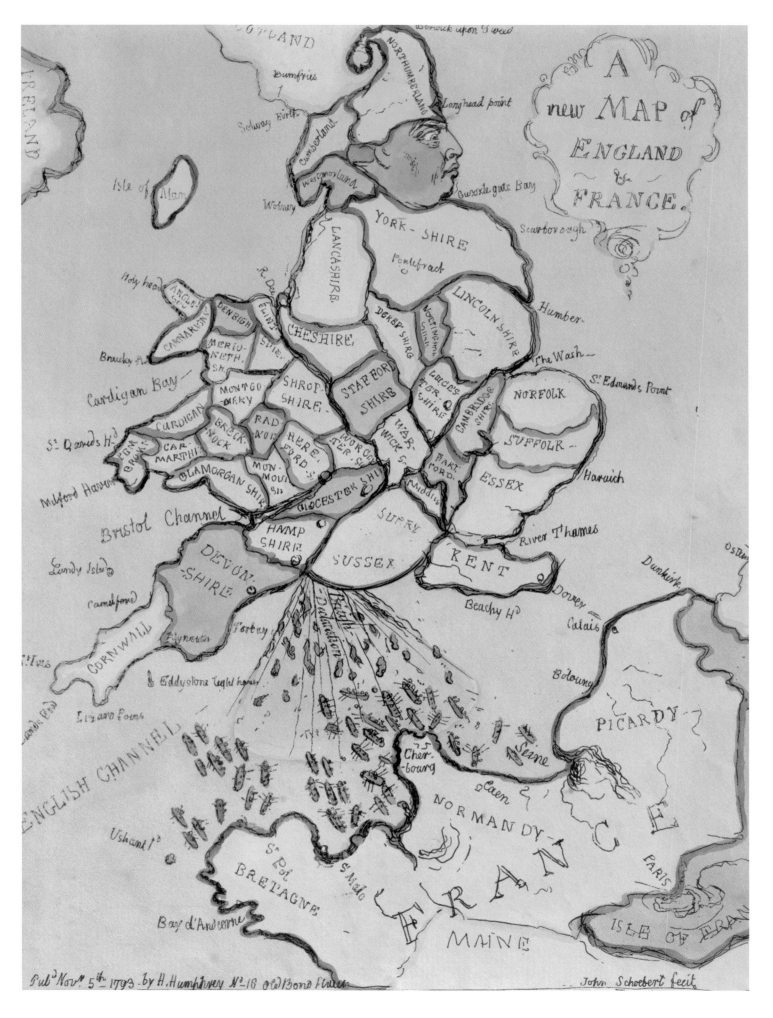

The French Invasion or John Bull Bombarding France With Bum-Boats, 1793, James Gillray

Courtesy New College, Oxford/
Bridgeman Art Library

This map hinges on the British fear of attack during the French Revolutionary wars as France seemed intent on exporting their regicide and revolutionary ardour across Europe—they had already invaded Austria and declared war on Britain. In the face of threat, the English retaliate with their favourite weapon, scatological invective. George III literally embodies England and vigorously excretes gunboats onto France. Hidden among the boats there is also a reference to The 'British Declaration'—a royal promise that the British occupied port of Toulon would be returned to the French, if and when they re-instated their monarchy.

After many years of unpopularity, the Hanoverian line of kings was finally beginning to achieve public acceptability and even inspire some enthusiasm. George III had had a disastrous start to his reign with the loss of the American colonies but the threat of revolution from across the channel and his partial recovery from debilitating illness (probably porphyria) appears to have allowed him to rebuild his popularity. Indeed, the fact that he is literally embodying England in this image is a strong message, an honour probably not bestowed to a British monarch since Elizabeth I. The map effectively stands for the State, but ignores any complexities involving Scotland or Ireland.

Britain in a Cloud, 1996, Rob Gandy
© Rob Gandy
In Rob Gandy's fleeting photography of cloud structures, we can see the general shape of Britain complete with Cornwall and Wales as prominent features. Scotland and North England are fainter while Ireland, though not accurately shaped, floats as it would on a map.

The photographer, Rob Gandy, claims to have taken the picture near Wadebridge in Cornwall on the morning of 3 August 1996. He noted: "It had been more 'solid' before I managed to get my camera, and as I watched, it slowly but surely broke up. Perhaps it was a portent of the effects of devolution following Tony Blair's election victory the following year."

RE-INVENTING ENGLAND

If satirists use maps as shorthand for their preoccupations, artists need to work the same material to such a degree that the result serves as more than just a one-liner. In such cases, contents and symbolism play dual roles in accruing a dialectic that does not become stale. The resonance is ultimately rewarding and leaves a charged view of England in its wake.

Maps have been a relatively new field of enquiry for artists. When Jasper Johns sought to signify America in his work of the 1960s, he appropriated the American flag and, more obliquely, a target board as universal objects for re-invention (rather than the highly recognisable map of America). Some of the earliest 'map works' can be dated back to the Surrealists *Map of the World* in 1929 and the later work of Guy Debord and the Situationist International in their re-invention of Paris in 1957. There was also some playful reworking of maps in artworks designed for Shell posters in the 1940s (for example, "Everyone Everywhere can be Sure of Shell" in 1944). The map as subject, however, really came in to its own in the 1980s and 90s when a wide range of artist latched onto, and began to explore, the possible layers of meaning inherent in them.

Tony Cragg's self-portrait in *Britain Seen from the North*, 1981, adopts a gloomy view of a land in economic turmoil but does so in the contradictorily primary colours of the detritus he has collected and assembled. His is an all-encomapssing view of Britain which, by the time of Jeremy Deller's *History of the World*, 1997, has dissipated and

fallen apart. This interconnected diagram of personal memories appears to represent a 'pained' Britain, a nation that has fragmented into its constituent parts and which, after the Scottish devolution a year later and Layla Curtis' *The United Kingdom*, 1999, is completely unrecognisable. In Curtis' collages geography has the disconcerting habit of being in the wrong place. Her series of *United Kingdoms* have abandoned their usual familiar content to be reformed from Japanese, American and European maps. As Britain becomes a more international, multicultural island—even while it reshapes itself into its constituent home nations—the population and the idea of 'home' is moving far more rapidly.

This conflation of 'nations' can be found again in Satomi Matoba's *Map of Utopia—The Japanese British Islands*, 2001, where the unfamiliar and familiar combine in promiscuous abandon. Ben Langland's and Nikki Bell's *Air Routes of Britain (Day) and (Night)*, 2000, further demonstrate that everyone is on the move, diaspora rife, as individuals leave traces of their restless traversing across the country and defy the separatist tendencies of those on the ground. It is easy to interpret such work in the context of some 20 years of political and national change. But perhaps such a reading is too simplistic. Deller's memories are personal, even if they do relate to public and semi-public events. Can they really be made to carry the weight of re-emergent national identity and the reworking of such ideas in a diverse and more culturally rich society than when Great Britain became great and the

United Kingdom united? While we desire to read the 'tea leaves' so to speak, to make legible the contradictions inherent in so many renderings of our national landscapes, like Grayson Perry's *Map of Englishman*, 2004, they often encompass too rich a brew to get the message straight. While cartography's explicit aim and purpose is to chart and measure, creative interventions such as these cleave open the map format to make way for the charting of abstract notions of self and nationhood.

From the very first attempts at mapping England, there is evidence of a desire to make the country legible politically, socially and geographically through the cartographic form, with these more interpretive efforts put forth by fine artists representing another level of association between self and country. Indeed, abstract notions of identity and place have always remained implicit in the mapping of England, and as ideas about what it means to be English have evolved, so too have representations of the land changed in equal measure.

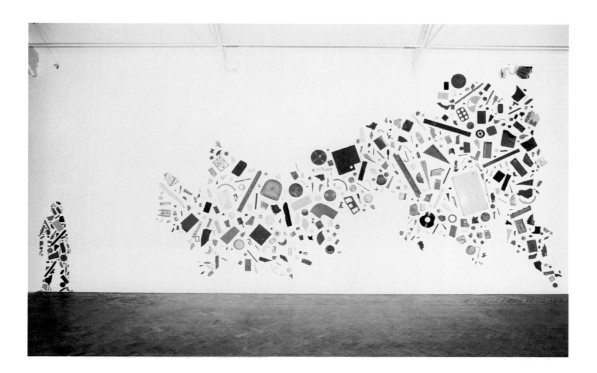

Above **Britain from the North, 1981, Tony Cragg**

Britain from the North is an unorthodox view of
the nation, including a portrait of the artist looking
sceptically at the country. As is typical of Tony
Cragg's work, it is created from found coloured
plastic pieces, broken and placed in a collage to
create new shapes. Cragg, by including himself in
his image, instantly makes a powerful subjective
comment on both mapping and his relationship to
the country mapped. In 1981, when this map was
created, Britain was plagued by social and economic
issues—mirrored in the fragmented pieces of plastic
that Cragg uses as his medium. Likewise, the map
is an allegory for the artist's increasing alienation
from his own country—fuelled by his relocation
to Germany.

Opposite **Sound Circle A walk on Dartmoor,
1990, Richard Long**

Courtesy the artist

Richard Long's powerful engagement with the
land, by walking on, recording and re-arranging the
landscape or ordering sticks, stones and mud on
the walls and floors of galleries signalled a new version
of engagement with topography, and indeed, maps
themselves. His work was only possible via the medium
of the map; it makes sense of his travels, connects the
viewer with the walking that they can only imagine
from graphic representation and takes them back to
their own physical experience of connecting maps to
the reality of a landscape. They are orienteering for the
imagination and re-introduced maps back into world
of art practice, from where they had been long absent.

Long's England is about experience, and although
entirely focussed on the countryside, it is a world
away from the romantic and mythologising versions of
landscape art and the picturesque imagination that had
become the preserve of the art form. Nonetheless, they
can provide a safe 'rural' experience for the 'urban' art
lover, not dissimilar to that of the armchair traveller
with an atlas of maps, comfortably perusing them in the
comfort of the home.

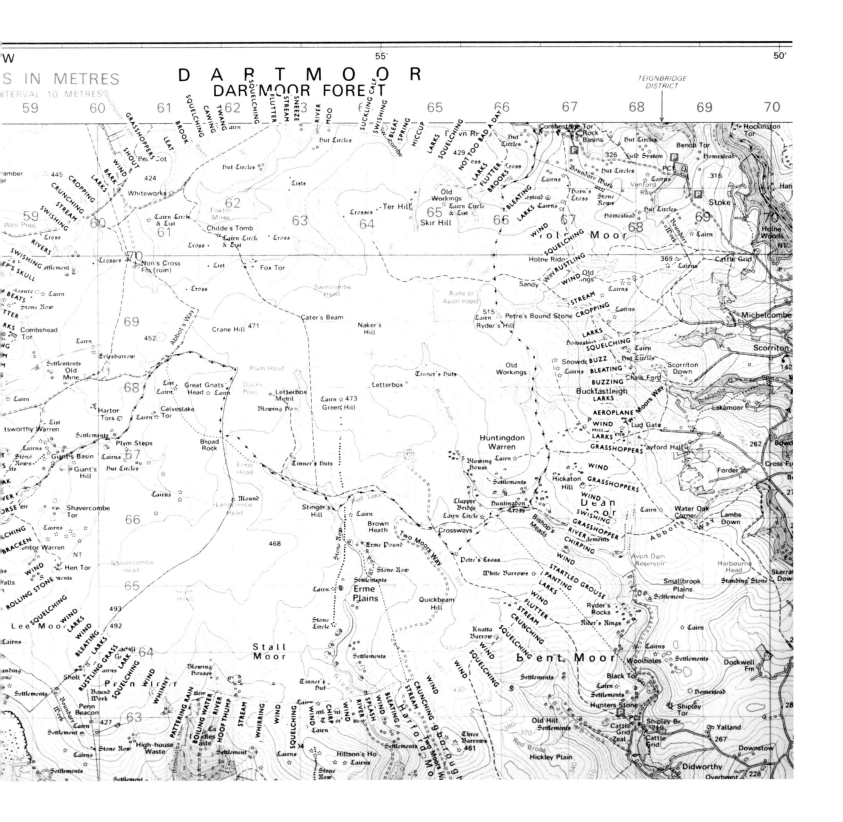

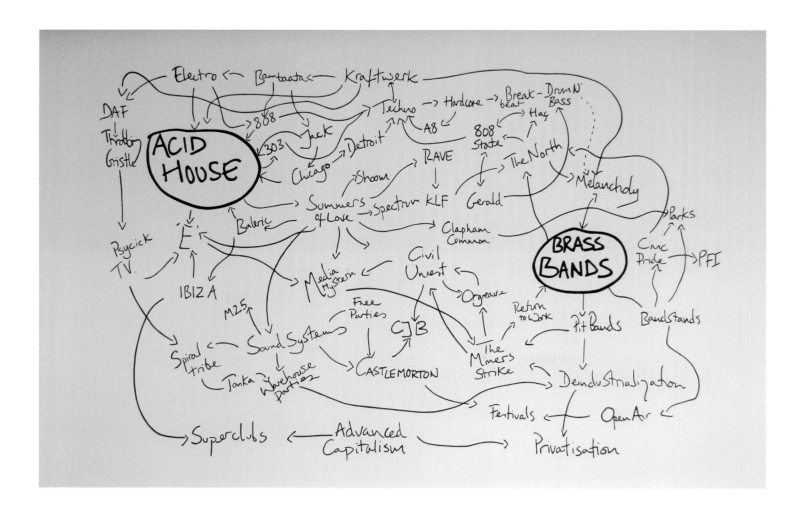

Above **The History of the World, 1998, Jeremy Deller**

Jeremy Deller created his *The History of the World* based on his 1996 sound-art piece, where he commissioned miners with brass bands to play acid house music. Deller then created this drawing to record his understanding of the links between the miners, brass bands and acid house. This represents a very different version of mapping, but expresses a freedom to use cartographic techniques to explore alternative aspects of England and the English imagination.

Opposite **Edit 2, 2000, Layla Curtis**
Courtesy the artist

Layla Curtis actively uses collages of real maps to re-invent the familiar, and to destabilise common assumptions about the relationship between maps and reality. We recognise this image immediately, but quickly register that it is somehow wrong. On closer inspection, the scale of local areas is distorted, and places do not appear in the geographical location we expect to find them. Curtis' work takes extreme liberties with what we think we know and plays

havoc with accepted certainties. The twenty-first century map, released from its representational role by the re-inventions of computer games and the like, should no longer be seen as entirely trustworthy, but possibly a more interesting guide to post-millennium England.

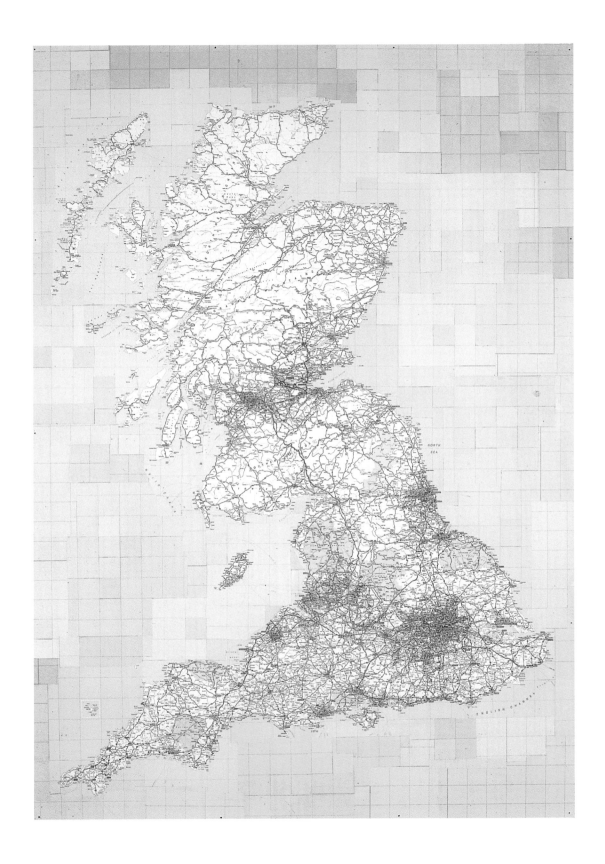

Above and opposite (detail) **The United Kingdom of Japan, 1999, Layla Curtis**
Courtesy the artist
In this work Layla Curtis continues to use collage to blur the boundaries between Scotland, England, Wales and Northern Ireland. This technique can also be seen in her other work from the same year, *United Kingdom*. Bringing a fictional combination of cities, towns and roads into the familiar form of the British Isles, Curtis encourages the viewer to question what features belong to which geographical location, in order to

fully engage with the illusion she has created. After graduating from Chelsea College of Art in 1999, Curtis lived in Japan for one year, where she began working on the hybrids for the maps that are shown here.

Overleaf **United Kingdom, 1999, Layla Curtis**
In *United Kingdom*, 1999, Layla Curtis explores ideas of national identity and questions of location through a map of Great Britain. To this end, she manipulates

the map and integrates Scottish and Welsh territory into English limits, and likewise replaces Scottish territory with that of the Welsh and English which forces Scotland to become an island on its own in the area where Ireland would be otherwise. *United Kingdom* questions how much we base our identities and culture on the geographical location we inhabit.

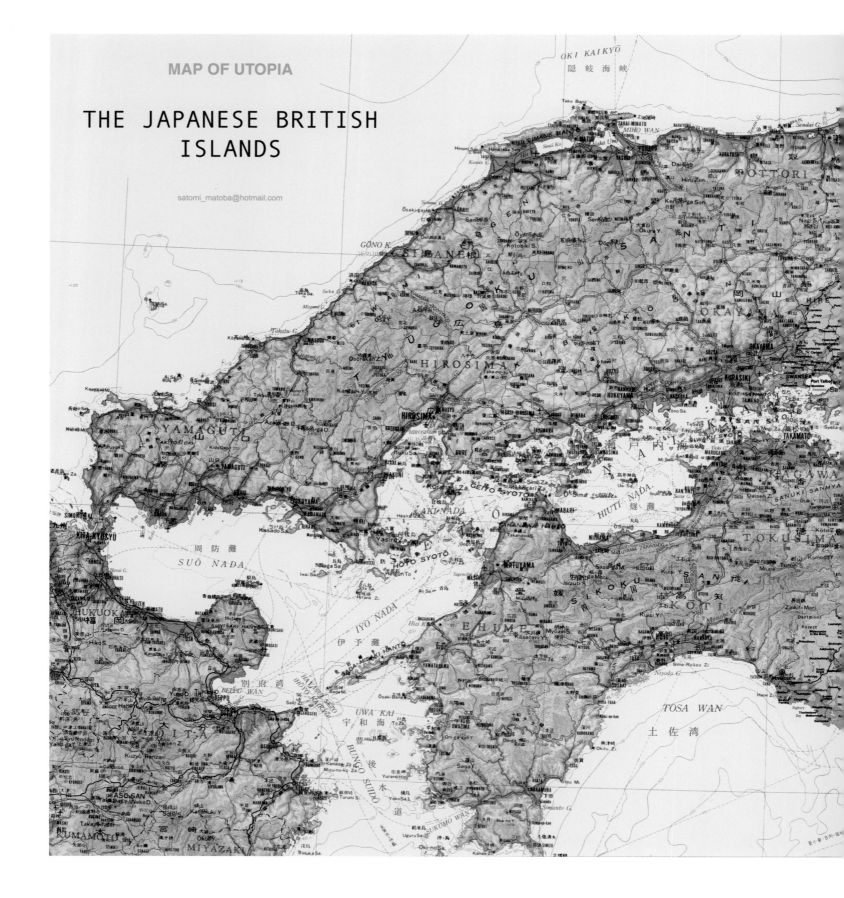

MAP OF UTOPIA

THE JAPANESE BRITISH
ISLANDS

satomi_matoba@hotmail.com

Japanese British Isles, 2001, **Satomi Matoba**
Courtesy the artist
Satomi Matoba looks at the universal language of
maps in her *Japanese British Isles*, using those signifiers
to create a map of a fictitious land. This work
is constructed of merged maps that describe an

eminently readable—though unfamiliar—territory,
full of allusions to the clashes between the east
and west and reflecting lives lived in a global world
instantly connected by networks that dissolve time
and distance between locations.

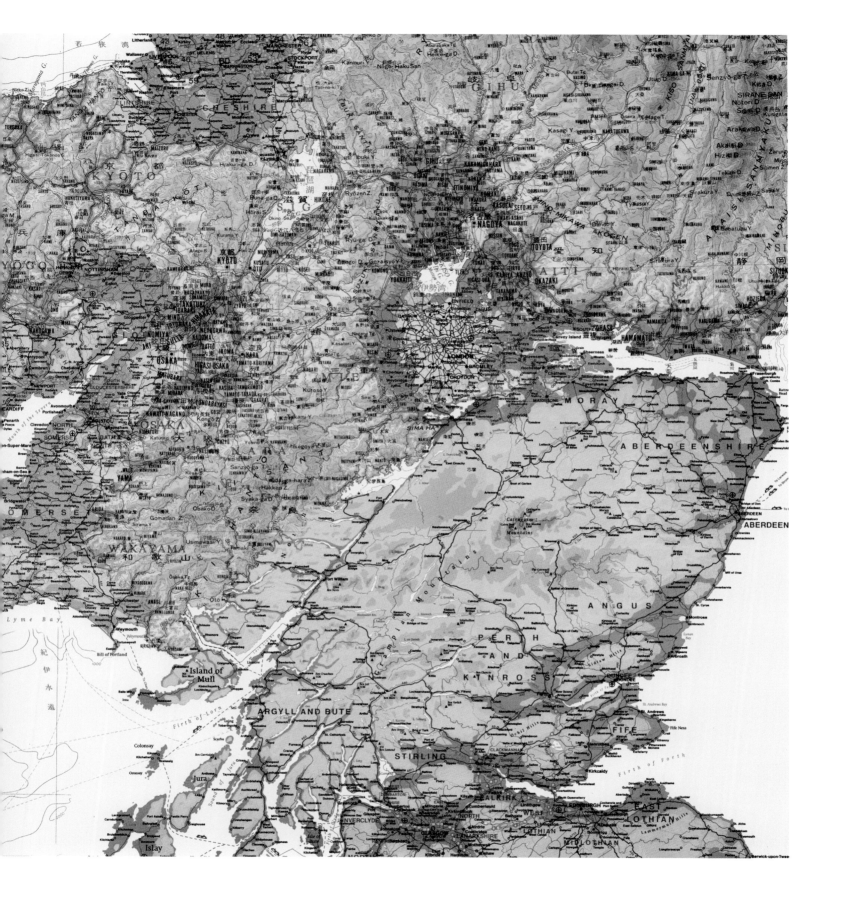

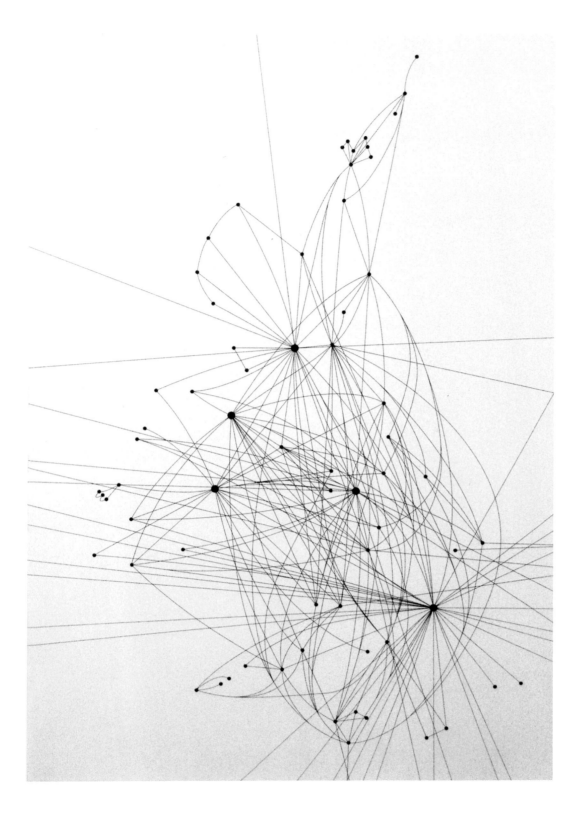

**Air Routes of Britain (Day and Night),
Langlands and Bell, 2000**

Courtesy Alan Cristea Gallery

Air Routes of Britain (Day and Night) is based on the
maps found in the back of in-flight magazines on

long-haul journeys, which usually depict a vague
presentation of the world with lines sporadically
crossing in various directions. Here, Langlands and
Bell take their departure from those approximate
representations of England, creating their own

version, seeking to question the intangible systems
that control our lives and the traces that we leave
behind as we travel from place to place.

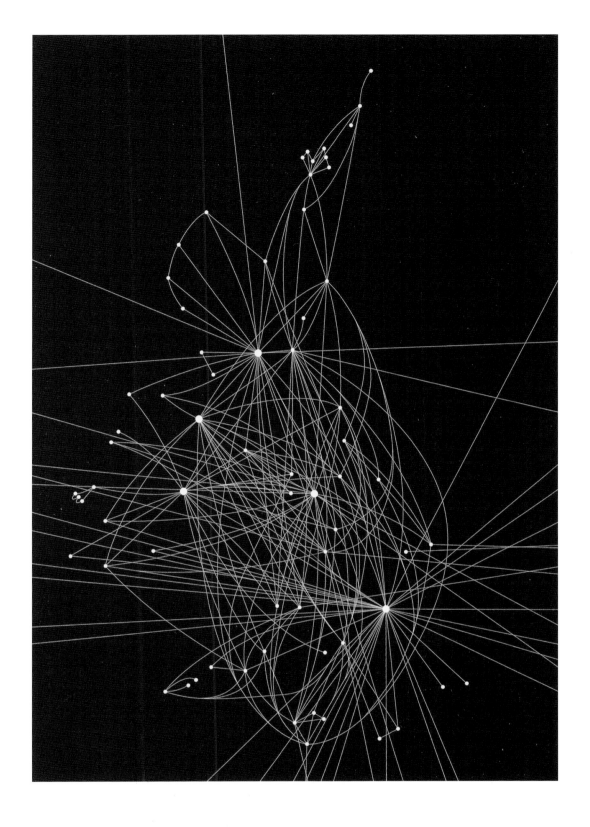

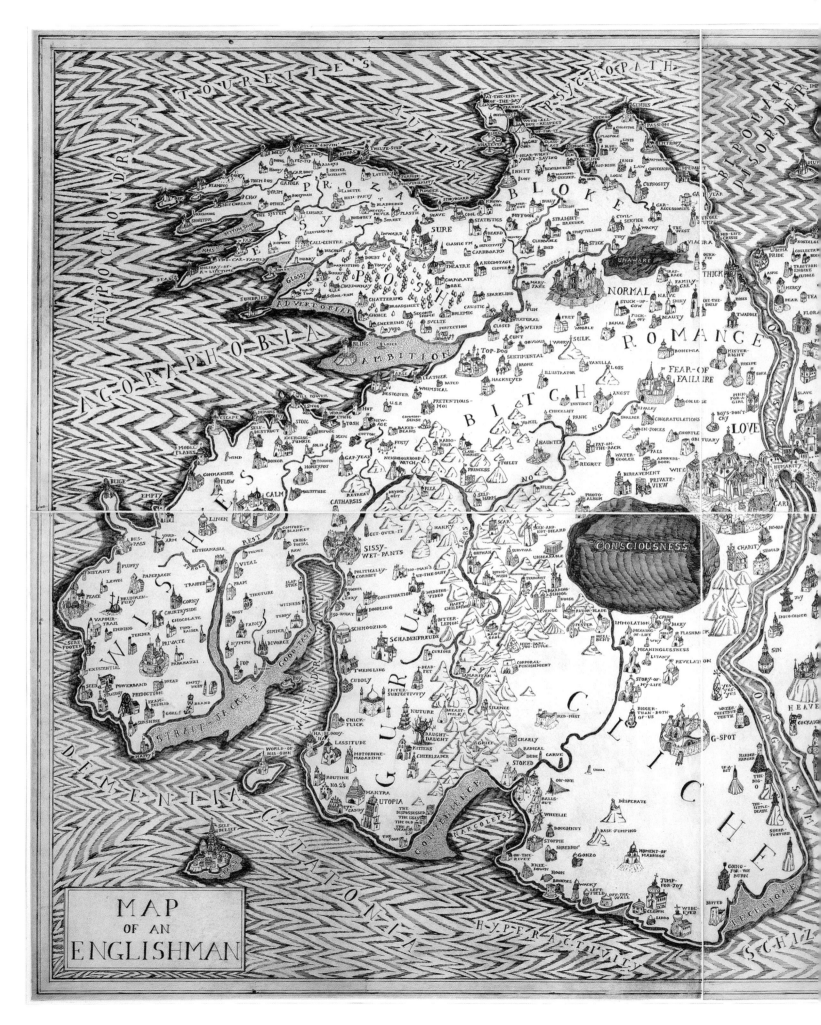

***Map of an Englishman*, Grayson Perry, 2004**

Grayson Perry's Tudor-style map explores his own psyche. In the centre of the etching lies an imaginary island, approximately in the shape of a brain. The island is surrounded by water teeming with an array of psychological disorders, such as schizophrenia, delirium, hypochondria and neurosis. The countryside inland is representative of Grayson's prejudices, fears and desires. There is anger, sex and dementia. Revealingly, cheerleaders and churches pepper the landscape and are labelled: suffocation, nappy, pride, bad day. Grayson's 'map' is reminiscent of phrenological study, because of its attempt at determining personality traits by mapping a physical entity. The viewer is able to enter into the artist's consciousness through the details of place names and areas.

ENDNOTES

INTRODUCTION

1   Febvre & Martin, *The Coming of the Book*, quoted by Benedict Anderson in *Imagined Communities*, London: Verso, 1983.

2   Quoted in *Some Conjectures about the Impact of Printing on Western Society and Thought: A Preliminary Report*, by Elizabeth L Eisenstein, University of Chicago Press, 1968.

MAPPING THE NATION

1   The Cotton Map is named after its seventeenth century owner, the antiquarian Sir Robert Cotton.

2   Kumar, Krishnan, *The Making of the English National Identity*, Cambridge: Cambridge University Press, 2003, p. 42.

3   Pevsner, Nikolaus, *The Englishness of English Art*, London: The Architectural Press, 1955, p. 83.

4   Agas, Ralph, *A Preparative to the Platting of Landes and Tenements for Surveigh*, 1596.

5   An open letter from the Privy Council, addressed to Justices of the Peace, 10 July 1576.

6   Delano-Smith, Catherine, and Roger JP Kain, *English Maps: A History*, London: The British Library, 1999.

7   Harvey, PDA, *Maps in Tudor England*, London: The Public Record Office and The British Library, 1993, p. 65.

8   Shakespeare, William, *Richard III*, circa 1595.

9   Delano-Smith and Kain, *English Maps: A History*.

10 Delano-Smith and Kain, *English Maps: A History*.

11 Close, Charles, *The Map of England*, London: Peter Davies, 1932.

USEFUL AND INFORMATIVE

1   Pepys, Samuel, *The Diary of Samuel Pepys*, 9 June 1667.

2   Close, Charles, *The Map of England*, p. 122.

3   Roy, William, *Military Antiquities of the Romans in North Britain*, The Society of Antiquaries, 1793.

4   Colls, Robert, *Identity of England*, Oxford: Oxford University Press, 2002, p.131.

5   "Mist hugged the moor, melted on the mountains" from Gawain and the Green Knight, Anon, fourteenth century.
    Lennon/McCartney, A Day in the Life, 1967.
    McEwan, Ian, *On Chesil Beach*, Jonathan Cape, 2007.

6   Forster, EM, *Howards End*, Edward Arnold, 1910, chapter 19.

7   Housman, AE, *A Shropshire Lad*, 1896.

8   Delano-Smith, C, and RJP Kain, *English Maps: A History*, London: The British Library, 1999, p. 90.

9   Quoted in Close, Charles, *The Map of England*, p. 59.

10 William Smith quoted in Winchester, Simon, *The Map that Changed the World*, Viking, 2001.

11 Osborne, Roger, *The Floating Egg Episodes in the Making of Geology*, London: Jonathan Cape, 1998.

12 Adam Sedgwick from his speech at the Geological Society presentation to William Smith.

13 "The earliest date at which the Government became in the redistribution and enclosure of communally held land in England was in 1604, when an act of parliament was obtained for the enclosure of Radpole, Dorset." Delano-Smith, C, and RJP Kain, *English Maps: A History*, p. 127.

14 Hoskins, WG, *The Making of the English Landscape*, Hodder and Stoughton, 1955.

15 Quoted in Delano-Smith and Kain, *English Maps: A History*, p. 125.

16 Eliot, George, *Middlemarch*, William Blackwood and Sons, 1874.

17 The shipping forecast areas, which have been broadcast four times daily on BBC Radio 4 since 1949. (There is a strict limit of 370 words to each broadcast.)

18 Tyndall, John, *Heat: A Mode of Motion*, London: Longmans, Green & Co, 1870.

19 Defoe, Daniel, *A Tour Through the Whole Island of Great Britain*, 1724–1726, London: Penguin, 1971.

20 Carroll, Lewis, *The Hunting of the Snark*, 1874.

21 Delano-Smith, C, and RJP Kain, *English Maps: A History*, 1999.

22 Ogilby, John, *Britannia*, Volume the First or an Illustration of the Kingdom of England and Dominion of Wales, 1675.

23 Ogilby, *Britannia*.

24 Ogilby, *Britannia*.

25 Coward, Noel, *Brief Encounter*, script, 1945.

IMAGINING ENGLAND

1   Milton, John, *Paradise Lost*, 1667.

2   Blake, William, "And did those Feet in Ancient Time", 1804.

3   Foxe, John, *Actes and Monuments of these Latter and Perillous Days, touching Matters of the Church*, John Day, 1663.

4   Peter Ackroyd in *Albion: The Origins of the English Imagination* (Chatto & Windus, 2002, p. 350) notes that "*Foxe's Book of Martyrs* thoroughly exemplifies the English tradition in its combination of improbable anecdote and broad theatricality".

5   Jones, Edwin, *The English Nation*, Sutton Publishing, 1998.

6   Colley, Linda, *Britons: Forging the Nation 1707–1837*, Yale University Press, 1992.

7   Ackroyd, Peter, *Blake*, London: Sinclair-Stevenson, 1995.

8   Benson, AC, *Land of Hope and Glory*, 1902.

9   Wilde, Oscar, *The Soul of Man Under Socialism*, 1891.

10 Carroll, Lewis, "The Hunting of the Snark", 1876.

ACKNOWLEDGEMENTS

This book is very much a follow-up to my previous book, *Mapping London: making sense of the city*, but although there is much shared history and many of the names of the cartographers are the same, the themes are very different. Whereas maps of London are required to interpret and bring order to a chaotic and ever-changing labyrinth of streets and the landmarks in them, the job of mapping England appears to be one of defining and crystallising the essence of a place that essentially has stayed the same since maps started to be drawn, even though the viewers have moved on and changed. Because the independent nation state of England only existed within its familiar geographic boundaries for the shortest of periods (between the fall of Calais and the ascension to the throne of James VI/I—essentially the Elizabethan period and no more), other means have been necessary to give identity to England and it is my belief that the practice of mapping the country has been one of the most significant and least recognised. In comparison, London had no need of maps to give it identity, even though it turned out that Henry Beck's Underground diagram became one of its most recognisable icons.

My first map of the country was a jigsaw puzzle, of the whole of the British Isles rather than just England. It was a puzzle that took several days of effort to complete during the many annual family holidays it went on, and parts of the country will be forever associated with the pictograms of various iconic places; the bridge of the Tyne, at Newcastle, or the Needles next to the Isle of Wight. Perhaps it sowed a lifelong belief that maps were more than route-finders and accurately scaled versions of the real thing and that, additionally, they had a great, and possibly greater, symbolic relevance. Or perhaps that realisation only came later. I am well aware that I am not alone in my fascination with maps, as so many people have told me how much they mean to them. What I never encountered, at least until I started reading and researching for this series of books, was any great public discussion of maps and their significance to the way so much of the rest of the graphic imagery we meet on a daily basis is dissected and analysed. I hope this book will go some way to encouraging more discussion of this under-sung medium, one that is nonetheless an essential part of all of our collective understanding of the world and particularly of the country we inhabit.

The writing of *Mapping England* has been essentially a solo effort, and all the errors and the exaggerations and simplifications of often complex issues are my own. I owe a great deal of thanks to my editor Blanche Craig and publisher Duncan McCorquodale at Black Dog Publishing, especially for keeping up the pressure in the nicest possible way, but greatest thanks are due to my family; Anne, Imogen, Mischa and Pascoe, for excusing me while I have been in pursuit, one way or another, of England.

Simon Foxell, August 2008

The editor wishes to thank Paul Teasdale for his invaluable help with picture research, Angela Torchio, Laura Barnicoat, Frances Copeman and Adam Salisbury for their insightful caption writing, as well as Nadine Monem for her support throughout the project. Thanks to Lizzie Keeling, Laura Mingozzi and Ana Estrougo for design assistance. Special thanks must go to Matthew Pull for his continual dedication to the project and for producing such a sensitively-designed volume.

BIBLIOGRAPHY

GENERAL REFERENCE

- Akerman, James, Robert Karrow and John McCarter eds, *Maps: Finding our Place in the World*, Chicago: The University of Chicago Press, 2007.
- Barber, Peter ed, *The Map Book*, London: Weidenfeld & Nicolson, 2005.
- Black, Jeremy, *Maps and Politics*, London: Reaktion Books, 1997.
- Buisseret, David, *The Mapmakers' Quest: Depicting New Worlds in Renaissance Europe*, Oxford: Oxford University Press, 2003.
- Delano-Smith, Catherine, and Roger Kain, *English Maps: A History*, London: The British Library, 1999.
- Dorling, Danny, *A New Social Atlas of Britain*, New Jersey: John Wiley & Sons, 1995.
- Foxell, Simon, *Mapping London: Making Sense of the City*, London: Black Dog Publishing, 2007.
- Federick, Charles, *The Map of England*, 1932.
- Harmon, Katherine, *You Are Here: Personal Geographies and Other Maps of the Imagination*, New York: Princeton Architectural Press, 2004.
- Harvey, PDA, *Maps in Tudor England*, London: The British Library & Public Record Office, 1993.
- Langton, J, and RJ Morris eds, *Atlas of Industrialising Britain 1780–1914*, London: Methuen, 1986.
- Millea, Nick, *The Gough Map: The Earliest Road Map of Great Britain*, Oxford: The Bodleian Library, 2007.
- Moreland, Carl, *Antique Maps*, London: Phaidon Press, 1989.
- Pickles, John, *A History of Spaces*, New York: Routledge, 2004.
- Roy, William, *Military Antiquities of the Romans in North Britain*, London: The Society of Antiquaries, 1793.
- Thrower, Norman JW, *Maps and Civilisation: Cartography in Culture and Society*, Chicago: University of Chicago Press, 1999.
- Tufte, Edward, *The Visual Display of Quantitative Information*, Cheshire: Graphics Press, 2001.
- Tufte, Edward, *Visual Explanations*, Cheshire: Graphics Press, 1997.
- van Roojin, P, *The Agile Rabbit Book of Historical and Curious Maps*, Amsterdam: Pepin Press, 2005.
- van Swaaij, Louise, and Jean Klare, *The Atlas of Experience*, London: Bloomsbury, 2000.
- Winchester, Simon, *The Map that Changed the World*, New York: Viking, 2001.
- Bull, G, *Thomas Milne's Land Use Map of London and Environs*, London: London Topographical Society, 1976.

HISTORY AND BIOGRAPHIES

- Ackroyd, Peter, *Blake*, London: Sinclair-Stevenson, 1995.
- Brown, Jane, *The Pursuit of Paradise: A Social History of Gardens and Gardening*, London: Harper Collins, 1999.
- Colley, Linda, *Britons: Forging the Nation 1707–1837*, London: Yale University Press, 2005.
- Clifton-Taylor, Alec, *The Pattern of English Building*, London: Faber and Faber, 1987.
- Davies, Norman, *The Isles: A History*, London: Macmillan, 1999.
- Foxe, J, *Actes and Monuments of these Latter and Perillous Days, touching Matters of the Church*, John Day, 1663.
- Hamblyn, Richard, *The Invention of Clouds*, London: Farrar, Straus and Giroux, 2001.
- Hobsbawm, Eric, *The Age of Revolution 1789–1848*, London: Weidenfeld & Nicolson, 1962.
- Hobsbawm, Eric, *The Age of Capital 1848–1875*, London: Weidenfeld & Nicolson, 1975.
- Hobsbawm, Eric, *The Age of Empire 1875–1914*, London: Weidenfeld & Nicolson, 1987.
- Hobsbawm, Eric, *Age of Extremes: The Short Twentieth Century 1914–1991*, London: Michael Joseph, 1994.
- Hoskins, WG, *The Making of the English Landscape*, London: Hodder and Stoughton, 1955.
- Hunt, Tristram, *Building Jerusalem: The Rise and Fall of the Victorian City*, London: Weidenfeld & Nicolson, 2004.
- Jardine, Lisa, *The Curious Life of Robert Hooke*, London: Harper Collins, 2003.
- Osborne, Roger, *The Floating Egg: Episodes in the Making of Geology*, London: Jonathan Cape, 1998.
- Porter, Roy, *English Society in the Eighteenth Century*, London: Penguin Books, 1991.
- Schaffer, Frank, *The New Town Story*, London: MacGibbon & Kee, 1970.
- Summerson, J, *Architecture in Britain 1530–1830*, London: Yale University Press, 1993.
- Tyndall, John, *Heat: A Mode of Motion*, Longmans, London: Green & Co, 1870.
- Welsh, Frank, *The Four Nations*, London: Harper Collins, 2002.

ENGLAND: IMAGE AND IDENTITY

- Ackroyd, Peter, *Albion: The Origins of the English Imagination*, London: Chatto & Windus, 2002.
- Anderson, Benedict, *Imagined Communities*, London: Verso, 1991.
- Balakrishnan, Gopal ed, *Mapping the Nation*, London: Verso, 1996.
- Colls, Robert, *Identity of England*, Oxford: Oxford University Press, 2002.
- Duffy, Maureen, *England: The Making of the Myth, From Stonehenge to Albert Square*, London: Fourth Estate, 2001.
- Girouard, Mark, *The English Town*, London: Yale University Press, 1990.
- Jones, Edward, *The English Nation: The Great Myth*, Stroud: Sutton Publishing, 1998.
- Kumar, Krishan, *The Making of the English National Identity*, Cambridge: Cambridge University Press, 2003.
- Paxman, Jeremy, *The English*, London: Michael Joseph, 1998.
- Perryman, M, *Imagined Nation: England after Britain*, London: Lawrence and Wishart, 2008.
- Pevsner, Nikolaus, *The Englishness of English Art*, London: The Architectural Press, 1956.
- Scruton, Roger, *England: An Elegy*, London: Chatto & Windus, 2000.

TRAVEL WRITING AND FICTION

- *Beowulf: a Verse Translation*, London: Penguin Books, 1973.
- Barnes, Julian, *England, England*, London: London: Jonathan Cape, 1998.
- Blake, William, *Selected Poems*, Oxford: Heinemann, 1957.
- Bryson, Bill, *Notes from a Small Island*, London: Doubleday, 1995.
- Carroll, Lewis, *The Hunting of the Snark (An Agony in Eight Fits)*, London: Macmillan, 1876.
- Chaucer, Geoffery, *The Canterbury Tales*, London: Penguin Books, 1951.
- Christie, Agatha, *Murder at the Vicarage*, London: Collins Crime Club, 1930.
- Cobbett, W, "Rural Rides", *The Political Register*, London, 1830.
- Davies, Pete, *This England*, London: Little, Brown and Company, 1997.
- Defoe, Daniel, *A Tour Through the Whole Island of Great Britain*, London: Yale University Press, 1991.
- Dickens, Charles, *The Posthumous Papers of the Pickwick Club*, Florida: Chapman & Hall, 1837.
- Eliot, George, *Middlemarch*, London: William Blackwood and Sons, 1871.
- Fielding, Henry, *The History of Tom Jones*, Andrew Millar, 1749.
- Ford, Madox, *The Good Soldier: A Tale of Passion*, Oxford: Oxford University Press, 1999.
- Forster, EM, *Howards End*, London: Penguin Books, 2003.
- Gaskell, Elizabeth, *North and South*, Florida: Chapman & Hall, 1855.
- Grahame, Keneth, *The Wind in the Willows*, London: Methuen, 1908.
- Hanley, Lynsey, *Estates: an Intimate History*, London: Granta Books, 2007.
- Hoban, Russell, *Riddley Walker*, London: Jonathan Cape, 1980.
- Ishiguro, Kazuo, *The Remains of the Day*, London: Faber and Faber, 1989.
- McEwan, Ian, *On Chesil Beach*, London: Vintage, 2008.
- Milton, John, *Paradise Lost*, 1667.
- More, Thomas, *Utopia*, London: Penguin Books, 1965.
- Naipaul, VS, *The Enigma of Arrival*, New York: Viking, 1987.
- Pepys, Samuel, *The Diary of Samuel Pepys*, 1667.
- Priestley, JB, *English Journey*, London: William Heinnemann, 1934.
- Ransome, Arthur, *Swallows and Amazons*, London: Jonathan Cape, 1930.
- Reavell, Cynthia, *Mr Benson Remembered in Rye, and the World of Tilling*, Rye: Martello Bookshop, 1984.
- Sawyer, Miranda, *Park and Ride: Adventures in Suburbia*, London: Little, Brown and Company, 1999.
- Stevenson, Robert Louis, *Treasure Island*, London: Penguin Books, 1999.

LIST OF ILLUSTRATIONS

INDEX

© 2008 Black Dog Publishing Limited,
London, UK, the artists and authors.
All rights reserved.

Black Dog Publishing Limited
10A Acton Street
London WC1X 9NG
United Kingdom

Tel: +44 (0) 20 7713 5097
Fax: +44 (0) 20 7713 8682
info@blackdogonline.com
www.blackdogonline.com

Edited by Blanche Craig and
Nadine Monem at BDP.
Designed by Matthew Pull at BDP.

ISBN 978 1 906155 51 3
British Library Cataloguing-in-Publication
Data. A CIP record for this book is available
from the British Library.

Black Dog Publishing Limited, London, UK,
is an environmentally responsible company.
Mapping England is printed on Fedrigoni
Symbol Freelife Satin, an environmentally-
friendly ECF (Elemental Chlorine Free)
woodfree paper with a high content of
selected preconsumer recycled material.

Printed in Italy.

architecture art design
fashion history photography
theory and things

black dog
publishing

london uk

www.blackdogonline.com